Thomas Southwell

The Seals and Whales of the British Seas

Thomas Southwell

The Seals and Whales of the British Seas

ISBN/EAN: 9783337328931

Printed in Europe, USA, Canada, Australia, Japan

Cover: Foto ©Andreas Hilbeck / pixelio.de

More available books at **www.hansebooks.com**

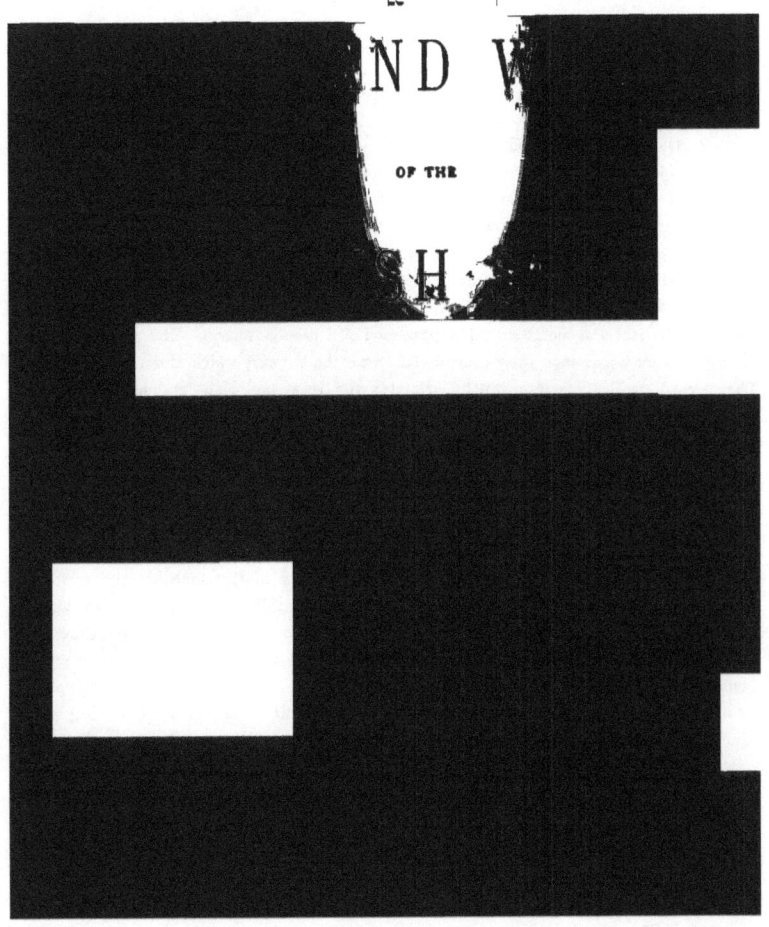

INTRODUCTION.

SEALS AND WHALES OF THE BRITISH SEAS.

ALTHOUGH at no period entirely neglected, as is apparent from the frequent reference to the subject by old authors, and from the known richness in species of the British Fauna, compared with that of the Continent of Europe, the study of the Marine Mammalia of the British Seas has, of late years, received more than usual attention, and the advance made in the knowledge of these creatures, has been rapid in proportion. Nor is it surprising that, to the inhabitants of a densely-peopled country like the British Isles, the terrestrial fauna of which must, of necessity, be very restricted and familiar, the study of the mammals frequenting its seas and shores should be possessed of a peculiar charm. The uncertainty and rarity of their occurrence, their exceptional forms, the mystery which shrouds their origin, heightened by the romance which surrounds the seas and high latitudes forming the chief home of so many species, must always render them objects of the greatest interest. Not only is this the case on the coast, but even in inland districts, whither—notably to London and Birmingham—Cetaceans have been brought, both living and dead, at great expense, and from long distances, to gratify the growing interest which has manifested itself, in these remarkable animals.

Under these circumstances it is surprising that no modern book, especially devoted to this subject, exists; those who would inform themselves must search out the scattered records dispersed in the publications of numerous Scientific Societies, or procure works, which, excellent as they may be, are much more comprehensive in scope, and too expensive to be within the reach of many into whose hands it is hoped this little book may come : the author has, therefore, striven to supply what is certainly a desideratum, viz., a cheap, plain, but, he hopes, trustworthy treatise on the Marine Mammalia of the British Seas. Originally published in the form of a series of papers in the pages of *Science Gossip*, the following account of the "Seals and Whales found in the British Seas" has been brought down to the present time, and much new matter added, not the least important of which is that devoted to the claims of the Atlantic Right-Whale to a place in the British fauna.

INTRODUCTION.

Doubtless, rare specimens are often lost to science for want of identification, and all those interested in their study have experienced the frequent disappointment which attends the bare announcement of "a Whale on shore:" in many instances no attempt is made to determine the species, in others it is evidently wrongly-named, or, although perhaps a more or less elaborate description may be given, not a single feature is indicated by which it may be identified.

One special object in reproducing these pages is to assist, by means of the most accurate figures which could be obtained, and short descriptions of the more important characters to be observed in each species, in determining those specimens which, from time to time, are landed by our fishermen, or cast dead upon the shore. Elaborate or technical descriptions have been carefully avoided, but short accounts of the habits and distribution, so far as known, of each species have been given, with the hope of interesting others in the study of this, even now, too-much-neglected branch of Natural History.

To the more advanced student the numerous references may be useful for indicating the sources whence detailed information of a more technical character is to be obtained.

The usefulness of this little manual, which pretends to no originality, but in the compilation of which no labour has been spared to insure accuracy, will, it is hoped, be greatly enhanced by the Illustrations; they were either engraved from original drawings, or copied from the most trustworthy sources (indicated in the text); several of them have since been adopted by the latest publications on the subject, both in England and America. For the use of 20 of the illustrations, out of a total of 29, the author is indebted to the kindness of Mr. David Bogue, who obligingly lent the blocks originally engraved for the papers in *Science Gossip*.

The author has to acknowledge, with many thanks, the kind assistance afforded him by MR. J. W. CLARK, Superintendent of the Museum of the University of Cambridge, and a recognized authority on the *Cetacea* and *Pinnipedia*. He, also, has to record the services, in behalf of this little work, rendered by one, who, beloved and lamented by many friends, has passed away since it has been in the press—the late MR. EDWARD RICHARD ALSTON. The wound inflicted by the early death of that amiable and promising naturalist is too fresh to admit of further reference.

Norwich, March 1881.

INDEX.

	PAGE
Atlantic Right-Whale	61
Balæna biscayensis	61
„ *mysticetus*	49
Balænoptera boops	70
„ *borealis* (Note)	128
„ *laticeps*	77
„ *musculus*	70
„ *rostrata*	78
„ *sibbaldii*	75
Beaked Whale	101
Beluga „	108
Bottle-head „	101
Bottle-nose Dolphin	124
Broad-fronted Beaked Whale	101
Cachelot	85
Cetacea	44
Cuvier's Whale	102
Cystophora cristata	24
Delphinapterus leucas	108
Delphinus acutus	125
„ *albirostris*	125
„ *deductor*	118
„ *delphis*	121
„ *globiceps*	118
„ *melas*	118
„ *phocæna*	120
„ *tursio*	124
Dolphin, Bottle-nosed	124
„ Common	121
„ Risso's	115
„ White-beaked	125
„ White-sided	125

	PAGE
Epiodon desmarestii	102
Globicephalus melas	118
Grampus, Common	113
„ Risso's	115
Grampus cuvieri	115
„ *griseus*	115
Greenland Right-Whale	49
Halichærus gryphus	28
Hump-backed Whale	69
Hyperoodon butzkopf	101
„ *latifrons*	101
„ *rostratum*	101
Lagenorhynchus acutus	125
„ *albirostris*	125
Megaptera longimana	69
Mesoplodon sowerbiensis	105
Monodon monoceros	106
Mystacoceti	49
Narwhal	106
Odontoceti	85
Orca gladiator	113
Phoca baikalensis	17
„ *discolor*	17
„ *grœnlandica*	21
„ *hispida*	14
„ *vitulina*	11
Phocæna communis	120
Physalus antiquorum	70
„ *latirostris*	75
Physeter macrocephalus	85

	PAGE		PAGE
Pilot Whale	118	*Trichechus rosmarus* .	32
Pinnipedia .	2	*Tursio truncatus* .	124
Porpoise .	120		
Pseudorca crassidens	114	Walrus . .	32
Risso's Grampus	115	Whale, Atlantic Right	61
		„ Beaked .	101
Rorqual, Common	70	„ Bottle-head	101
„ Lesser .	78	„ Broad-fronted .	101
„ Rudolphi's .	77	„ Cuvier's .	102
„ „ (Note)	128	„ Greenland Right	49
„ Sibbald's .	75	„ Humpbacked	69
Rorqualus minor .	78	„ Pilot . .	118
Seal, Common .	11	„ Sowerby's .	105
„ Greenland	21	„ Sperm .	85
„ Grey . . .	28	„ White .	108
„ Hooded, or Bladder-nosed	24	White-sided Dolphin .	125
„ Ringed, or Marbled .	14	White-beaked Dolphin	125
Sibbaldius borealis	75		
Sowerby's Whale	105	Ziphioid Whales	98
Sperm Whale	85	*Ziphius cavirostris*	102

ERRATA.

Page 77, bottom line, for *Physalis* read *Physalus*.
„ 126, for *alberostris* read *albirostris*.

LIST OF ILLUSTRATIONS.

		PAGE
Figure	1.—Hind Flippers of Ringed Seal	2
,,	2.—Skeleton of Seal	12
,,	3.—Ringed or Marbled Seal	15
,,	4.—Greenland Seal	20
,,	5.—Hooded Seal	. 25
,,	6.—Grey Seal	29
,,	7.—Walrus	33
,,	8.—*Vacca Marina*	37
,,	9.—Head of Walrus	. 39
,,	10.—Sea Horse (after Cook)	41
,,	11.—Section of Skull of Whalebone Whale	46
,,	12.—Greenland Right-Whale	51
,,	13.—Atlantic Right-Whale	. 60
,,	14.—Common Rorqual	71
,,	15.—Lesser Rorqual	. 80
,,	16.—Sperm Whale	84
,,	17.—Chair in Great Yarmouth Church	. 87
,,	18.—Back View of ditto, ditto	87
,,	19.—Skeleton of Sperm Whale	. 88
,,	20.—Skull of Ditto	90
,,	21.—Head of Sowerby's Whale	. 104
,,	22.—Beluga, caught by the tail	109
,,	23.—Grampus	. 112
,,	24.—*Pseudorca crassidens*	114
,,	25.—Risso's Dolphin	. 116
,,	26.—Pilot Whale	118
,,	27.—Common Dolphin	. 122
,,	28.—Bottle-nosed Dolphin	124
,,	29.—White-beaked Dolphin	. 126

Table of British Cetacea		48
,,	Differences of British Mystacoceti	. 82

SEALS AND WHALES

OF THE

BRITISH SEAS.

THE two great groups of Marine Mammals known as *Pinnipedia* and *Cetacea*, although widely separated from each other zoologically, naturally present themselves to us side by side as inhabiting the same regions; the facilities for studying the one are also equally favourable for obtaining a knowledge of the other. It is remarkable that in few groups of the animal world, until recently, has so much confusion existed as in the Seals and Whales. This has, of late years, through the labours of European and American naturalists, to some extent been remedied, although very much still remains to be done, the literature of the subject being still so scattered, that much of it is inaccessible to the ordinary student. The arrangement and nomenclature adopted in the following short account of the Seals and Whales inhabiting or occurring in the seas, or on the shores, surrounding the British Islands, is that used by Mr. Alston in the second edition of Bell's 'British Quadrupeds.'

PINNIPEDIA.

The *Pinnipedia* (fin-footed) forms a well-marked sub-order of the Carnivora, and may be divided into three distinct families—the *Phocidæ*, or true Seals; the *Trichechidæ*, represented by one species only—the Walrus; and the *Otariidæ*, or Eared Seals.

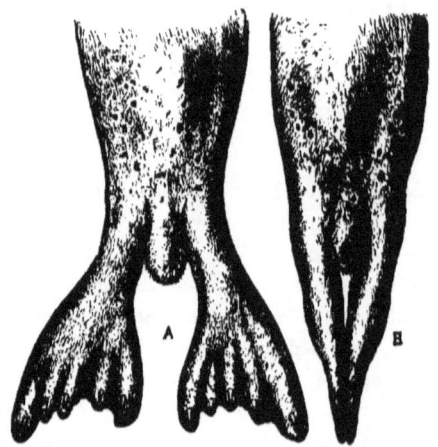

Fig. 1. HIND FLIPPERS OF RINGED SEAL *(after Murie).*
A, opened out; B, closed.

The *Phocidæ* are found both in the Northern and Southern hemispheres, most plentifully in the cold regions, but extending into the temperate seas; in the Northern hemisphere they are found as far south as 40° N. latitude; two species, however, are said to be sub-tropical. The true Seals may readily be distinguished by the absence of external ears, and the position of the posterior limbs, which are not adapted for progression on land, but admirably suited for propelling the animal through the element in which it obtains its sustenance. These limbs are directed backwards, and compressed laterally, the soles of the flippers being turned inwards, and are only free from the ankle-joints. (Fig. 1). Like the whole group, the Seals are carnivorous. Five species are believed to have occurred on our shores.

The family of *Trichechidæ* is limited to one genus, and that consisting of only one species, the Walrus or Morse, which is essentially Arctic in its habitat, and on our coasts can only be regarded as a very rare and accidental straggler; in this animal there is no external ear; its limbs are adapted for raising the body from the ground, thus enabling it to progress by their means upon dry land.

The third family, *Otariidæ*, consists of several genera and species (according to Gray); they are distinguished from both *Phocidæ* and *Trichechus* by the presence of external ear-conchs, and from the former by the structure of their limbs, which are free and adapted for progression upon land, where at a certain season they take up their abode for a considerable period. Dr. Pettigrew also points out that the fore-feet are hardly used by the true Seals as means of propulsion in the water, whereas in the Eared Seals they form the chief organs used for that purpose, and in the Walrus all four limbs are employed. The Eared Seals inhabit the lonely shores and islands of the Pacific Ocean and South Seas, where they are hunted for their skins; the beautiful "seal-skin" of commerce, so much prized for its lustre and softness, being the dyed and prepared under-fur of some members of this family. The *Otariidæ* are not represented in our fauna.

The true Seals spend most of their time in the water, but visit the shore or ice to bask in the sun or bring forth their young; this last takes place early in the summer, and it is seldom that more than one is produced at a birth. Some species enter the water almost immediately after birth, but others are two or three weeks before they leave the ice, quitting it at first very unwillingly, but soon becoming expert at swimming and diving. The power of the Seal to remain beneath the water for lengthened periods Dr. Wallace* believes to be acquired rather than structural. Their food consists of crustacea and fish, with an occasional sea-bird. Some species are migratory in their habits. In disposition they are usually timid and gentle, and capable of attachment, when in confinement, to those who feed and attend them. The Bladder-nose and Grey Seals, however, appear to be exceptions to this rule; the former is said to be fierce and vindictive, rather courting than fleeing from danger, and altogether a formidable opponent. Their great affection for their young is made use of by the sealers for their destruction.

Although Seals are not found in sufficient numbers round our own coast to be of any commercial value, in the Northern Seas, where they congregate in vast numbers at the breeding season, the seal-fishery is of great importance as a branch of industry, and finds employment for a large number of vessels and men, both from this country and from the ports of Northern Europe. In the Greenland seal-fishery the Norwegian whalers had in 1874 sixteen steamers and nineteen sailing-ships, with an aggregate tonnage of 9,000 tons, manned by 1,600 sailors, and in the three years ending 1874 they killed 142,500 young Seals and 128,000 old ones, notwithstanding which the balance-sheet of the three years showed only a small profit on the steamers and a large loss on the sailing vessels.† An official return issued by Messrs. David Bruce and Co., of Dundee, shows that in the season of 1879, eleven Dundee ships and five from Peterhead, were engaged in the Greenland seal-trade; the total catch of these

---

* Dr. Robert Brown on the 'Seals of Greenland.' Reprinted, with additions, in the 'Manual and Instructions for the Arctic Expedition, 1875,' from the *Proc. Zool. Soc.*, 1868, pp. 405-440.

† *Land and Water*, August 26th, 1875.

sixteen ships was 35,044 Seals; four ships from Dundee visited Newfoundland and captured 70,355 Seals, making a total for the British ships alone of 105,399 Seals, exclusive of those wounded and lost, or otherwise destroyed. These produced 1280 tons of oil, worth about £25 per ton, or £32,000, exclusive of skins, which sell for about 5s. each. The majority of the Norwegian vessels also bring their cargoes to this country. Captain David Gray informs me that the seal-fishery was commenced from the Port of Peterhead, in the year 1819, since which time to the close of the season of 1879, the large number of 1,673,052 Seals have been taken by the vessels belonging to that port. The Dundee vessels did not take part in the seal-fishery till the year 1860, but have from that time to 1879 taken 917,278 Seals. This total is greatly swollen by the results of the Newfoundland fishery; four Dundee vessels in 1879 took 70,355 Seals in Newfoundland, whereas, in the same season, eleven Dundee and five Peterhead vessels took only 35,044 Seals in the Greenland fishery. The Dundee ships, after the Newfoundland fishery is ended, generally land their oil and skins at St. John's, and proceed on their whaling voyage to Greenland and Davis' Straits.

Dr. Wallace [*] estimates the annual produce of the Greenland Seal-fishery alone at the sum of £116,000; the bulk of the seals taken are the Harp-. Seal *(Phoca grœnlandica)*.

Several attempts had been made to establish a seal-fishery at Newfoundland, from the port of Dundee, but with small success till the year 1876: in that year Messrs. Alexander Stephen and Son secured premises at St. John's, and sent out two vessels to be manned chiefly by a Newfoundland crew; the result was a great success, and this firm has since prosecuted the fishing with very satisfactory results. The Dundee Seal and Whale Fishing Company have also three steamers in the trade, in addition to those engaged at

---

[*] Dr. Brown's 'Seals of Greenland,' *Proc. Zool. Soc.*, June, 1868, reprinted in the 'Arctic Manual,' p. 67.

the Greenland fishery. Mr. David Bruce, of Dundee, to whom I am indebted for the above particulars, informs me that the season of 1880 was a failure in the Newfoundland fishery, and that out of a fleet of twenty-four steamers, not more than six of them would pay their expenses.

Mr. J. A. Allen* gives an interesting account of the rise and progress of the Newfoundland fishery, which he characterises as "the sealing-ground, *par excellence*, of the world, twice as many Seals being taken here by the Newfoundland fleet alone as by the combined sealing-fleets of Great Britain, Germany, and Norway, in the icy seas about Jan Mayen, or the so-called 'Greenland Sea' of the whalemen and sealers." So early as 1721, thousands of "sea-wolves" were killed in the Gulf of St. Lawrence, but, according to Mr. Michael Carroll, of Bonavista, Newfoundland, in his account of the 'Seal and Herring Fisheries of Newfoundland,' published in 1873, as quoted by Mr. Allen, it was not till the year 1763 that the seal-fishery was regularly prosecuted there by vessels specially equipped for the purpose. The trade, however, rapidly assumed importance, and in 1807 thirty vessels from Newfoundland alone were engaged in it. In 1834 the Newfoundland fleet had increased to three hundred and seventy-five, besides a considerable number of vessels from Nova Scotia and the Magdalen Islands; in 1857 the number of vessels employed appears to have reached its maximum, exceeding three hundred and seventy, whilst the catch of Seals was estimated at 500,000. About the year 1866, steamships were first introduced, and have ever since been increasingly employed; the result has been a steady decrease in the number of vessels, which, in 1871, were reduced to one hundred and forty-six sailing vessels and fifteen steamers, or less than one-half, but the number of Seals taken annually, up to 1873, appears to have remained about the same,

---

* 'History of North American Pinnipeds,' by Joel Asaph Allen. U.S. Geological and Geographical Survey of the Territories, Miscellaneous Publications, No. 12, Washington Government Printing Office, 1880.

and, notwithstanding the enormous destruction of these creatures, which takes place every season on the Newfoundland sealing grounds, many thousands of which, from the wasteful methods employed in their capture, are never accounted for, Mr. Carroll is still of opinion that up to the year 1873, their numbers were actually on the increase: this can hardly continue much longer to be the case.

I will only mention one of the methods employed by the Newfoundland sealers, which must eventually be attended with the most disastrous effects. This mode is technically called "panning." Mr. Carroll, writing in 1871 says, "No greater injury can possibly be done to the seal-fishery than that of bulking Seals on pans of ice by crews of ice-hunters. Thousands of Seals are killed and bulked, and never seen afterwards. When the men come up with a large number of old and young Seals, that cannot get into the water, owing to the ice being in one solid jam, they drive them together, selecting a pan surrounded with rafted ice, on which thousands of Seals are placed one over the other, perhaps fifteen feet deep. A certain number of men is picked out by the ship-master to pelt and put on board the bulked Seals, whilst other men are sent to kill more. It often happens that the men are obliged to go from one to ten miles, before they come up with the Seals again, and very often the men pile from five hundred to two thousand in each bulk, which bulks are from one to two miles apart; care is also taken that flags are stuck up as a guide to direct the men where to find such bulked Seals. So uncertain is the weather, and precarious the shifting about of the ice, as well as heavy falls of snow and drift, that very often such bulked Seals are never seen again by the men that killed and bulked them, as the vessels and steamships are frequently driven by gales of wind far out of sight or reach of them, and frequently wheeled or driven into another spot, when the men again commence killing and bulking as before. In many instances it has happened that the crews of vessels, as well as the crews of steamships, have killed and bulked twice their load. No doubt Seals that are bulked are often picked up by the crews of other vessels, but such is the law, that as long as the flags are erected upon the bulks, and the vessel or steamship is in sight, no man can take them, notwithstanding the vessel's or steamship's men that bulked them may be ten miles away from them, whilst another vessel may be driven within a quarter of a mile of thousands of bulked Seals, but, owing to the law, dare not take them." The skins, if left, are also liable to injury by the frost or sun, or by the capsizing of the pan they may be totally lost. In the spring of 1872, some five thousand Seals, obtained to the westward of Bonavista, by the inhabitants of that place, were heaped upon the ice. " There were thirteen flags to be seen in the morning over bulked Seals, and when the drift ice struck the land in the evening, only six of the flags were visible, the ice having rafted over both flags and Seals. Some days after, when the ice moved off from the shore, several bulks of Seals were found, but in such a putrid state that they could not be handled.* Comment upon the consequences which must speedily result from such lamentable waste of life is needless.

Nor, until very recently, was the seal-fishery in the Greenland Seas prosecuted with any greater regard to humanity or economy. " Supposing the sealing prosecuted with the same vigour as at present," says Dr. Brown, " I have little hesitation in stating that before thirty years shall have passed away, the seal-fishery, as a source of commercial revenue, will have come to a close, and the progeny of the immense number of Seals now swimming about in Greenland waters will number but comparatively few." Dr. Brown's remarks were written in the year 1868, and the prediction is already virtually fulfilled: a report, giving an account of the success of the Dundee vessels employed in the Newfoundland seal-fishery in 1877, after stating that 39,000

---

* 'Seal and Herring Fisheries of Newfoundland,' pp. 32-34, as quoted by Allen, *l. c.*, pp. 551-3.

Seals were said to have been captured by two vessels, concludes thus:—
" Previously all Dundee vessels were employed at the *Greenland* seal-fishing, but Captain Adams has for some years been of opinion that *that ground is practically used up*, and hence his visit to Newfoundland."

I will spare the reader, as much as possible, a repetition of the horrors of this cruel trade, and make only a single quotation from a letter written by an old and experienced sealer, Captain David Gray, of the steamship *Eclipse*. He says that five ships in 1873 shot among the old Seals for four days until the pack was utterly ruined. "I suppose," he continues, "about 10,000 old Seals had been taken. Add 20 per cent. for Seals mortally wounded and lost, gives an aggregate of 12,000 old ones; add 12,000 young ones which died of starvation (their parents being killed before the young ones were of any value or able to shift for themselves), gives 24,000. . . . The whole of the young brood was destroyed, and had these Seals been left alone for eight or ten days, I am quite within the mark when I say that, instead of only taking 300 tons of oil out of them, 1,500 could as easily have been got, and that without touching an old one."* So great are the cruelties perpetrated by the crews of the sealers, that even the men themselves, hardened as they are, sicken at the work, and cry shame that the law does not put a stop to them. Let anybody who cares to know what fearful cruelties man is capable of perpetrating for gain, read Captain Gray's letter. As a remedy for this waste of life (of course its cruelties can only be modified) Captain Gray suggested that the ships should be kept from sailing before the 25th of March, about a month later than they then started; they would then not reach the fishery and find the young Seals until they were sufficiently grown to be worth killing, and the frightful waste of life which occurred from the destruction of the old Seals before the young

---

* *Land and Water*, May 9th, 1874.

ones were able to shift for themselves, resulting in the death from starvation of the whole brood, thus be put a stop to.

With this object in view, an Act was passed in 1875, in which the Foreign States interested concurred, prohibiting the killing of the Seals before the 3rd of April in each year; from some misunderstanding this Act was not enforced in the season of 1876, but in 1877 it was rigidly observed by the ships of all nations engaged in the fishery. The result of the season's fishing was very unsatisfactory, owing to the absence of the large bodies of Seals which formerly were met with. Captain Gray, after three years' experience of the operation of this Act, considers that the fishing still opens too early,* and that an additional three days are necessary to enable the young Seals to arrive at their best, and prevent the useless slaughter of the old ones, which are getting thin from being suckled. He is of opinion that, since the introduction of the close time, the Greenland Seals are not diminishing quite so rapidly as they were, but that the restriction has not been in operation long enough to form a very accurate opinion.

The Walrus is even more rapidly and surely becoming exterminated than the Seal; it has become extinct from station after station, and but for its ice-loving habits, which render its present strongholds always difficult and sometimes impossible of access, it would now probably, like Steller's Rhytina, have to be spoken of in the past tense.

---

* Great diversity of opinion, however, exists upon this point, the Dundee sealers considering that the fishery should open a few days earlier, and that a time should be fixed for its closing, in order that too great a number of the old Seals may not be shot. The young Seals grow with great rapidity, and even a few hours make a marked difference in their condition; it seems, therefore, of the greatest importance that a time should be fixed for the opening of the fishery, which will ensure the young animals being in as forward a condition as possible, and that the nursing mother should be spared. It is said, also, that, in consequence of the number of females killed while nursing, the old dog Seals are vastly more numerous than the females, and that positive good is accomplished by some of them being killed off. One opinion, however, seems universal, which is, that not much good has resulted, at present, from the close time.

THE COMMON SEAL.

This species, *Phoca vitulina*, of Linnæus, is, *par excellence*, the COMMON SEAL of the British waters. It is found, although in greatly reduced numbers, on unfrequented shores and sands, from the Orkney and Shetland Islands, where it most abounds, to Cornwall, occasionally ascending estuaries and rivers for a considerable distance, but never quitting the immediate vicinity of the water. According to Bell, it occurs on both sides the North Atlantic, and is common in Spitzbergen, Greenland, and Davis's Straits; also Northern Russia, Scandinavia, Holland, and France, and is said to occur occasionally in the Mediterranean.* It figures largely in the returns of the Danish and Greenland fishery, where the number killed annually of this species and *Ph. hispida* is estimated by Dr. Brown at about 70,000.

Low, who died in 1795, says in the 'Fauna Oncadensis,' "A ship commonly goes from this place once a-year to Soliskerry, and seldom returns without 200 or 300 Seals;" these they killed by landing on the rock, and knocking them on the head. He also says that in North Ronaldsha they take them for the purpose of eating, and that the inhabitants say "they make good ham." Though at present far less numerous than formerly, it is still abundant in the unfrequented bays and sounds of the Orkney and Shetland Islands; also, on the Hebrides. On the mainland, Mr. Alston ('Fauna of Scot.' *Proc. Glasgow Nat. Hist. Soc.*) says it is found in all localities where it is free from intrusion, especially on the North and West shores; it is also common on some parts of the Irish Coast. In Wales it is not un-

---

* The Seal of the Caspian Sea was described as a variety of *Ph. vitulina*, by Pallas, and as a distinct species, by Nilsson, under the name of *Ph. caspica*. It is, however, notwithstanding its abundance, very little known, and may, probably, prove to be more nearly allied to the next species. The yearly average of this species taken in for the six years ending 1872, as given by Schultz, is 130,000.

common, and on the Cornish, and some few other favoured localities of the English coast it is still well known; on other parts of our shores it is decidedly rare. In the great estuary between the Norfolk and Lincolnshire coasts, called the "Wash," this species frequents the sand-banks left dry at low water, and, doubtless, many young ones are produced there annually. At birth, which takes place about the month of June, the young Seal is covered with a coat of white woolly hair, which is shed in parturition, or shortly after, and the young one takes to the water when only a few hours old. Mr. Bartlett gives an account of the birth of a young one (at the time

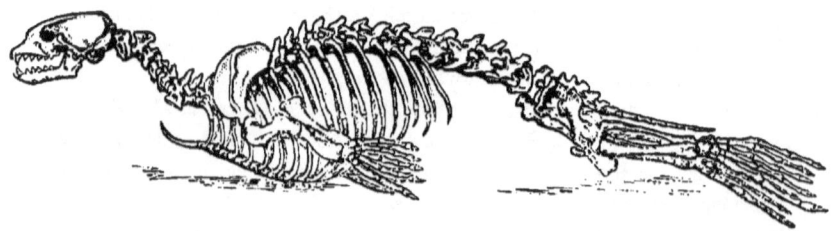

Fig. 2. SKELETON OF SEAL.

believed to be *Ph. hispida*) in the Zoological Gardens,* and states that it completely divested itself of its coat of fur and hair in a few minutes, and was swimming and diving about within three hours of its birth; its mother turned on her side to let it suck, and its voice was a low, soft "ba." The first coat is not shed so quickly in some species, nor do they all take to the water at so early an age; as, for example, *Ph. grœnlandica*, which is two or three weeks before it leaves the ice.

---

* *Proc. Zool. Soc.*, 1868, p. 402.

The total length of the adult is about 4 to 5 feet, and its coat is generally of a yellowish colour, thickly spotted with black on the back and upper parts, but less distinctly so on the sides. The under parts are a bright silvery hue; there is, however, considerable variety in colour and in the distinctness of the spots. This species is readily domesticated, and displays great intelligence, and even affection for those who feed and tend it. Almost everybody must have been struck with the docility displayed by the Seals which are occasionally exhibited as "talking fish." At the Zoological Gardens and at the Brighton and other Aquaria, where they are a never-failing source of attraction, their graceful movements in their confined homes cannot fail to excite admiration. Swimming silently and swiftly along, the animal threads with the greatest accuracy the intricacies of its narrow pond, assuming every possible attitude, and turning over and over in its course, as much at ease when swimming on its back as in its usual position. When, tired with this exercise, it comes to the edge of its pond and raises itself out of the water, its rounded head, and bright, full black eyes have something almost human in their expression, and the fabled "mermaid" seems a reality; but when once it leaves the water, it is clearly seen that it is no longer in the element in which it is destined to live and move, for its motions are laboured and awkward in the extreme. It throws itself along, first on one side and then on the other, just as a man tightly sewn in a sack would do, but, notwithstanding its clumsiness, contrives to make considerable progress.

This species may be distinguished by the arrangement of its molar teeth, which are placed obliquely along either side of the jaw, not in a line with each other. It has been said that this is only a characteristic of youth, and that the peculiar arrangement disappears "before the skull attains its maximum size." In the second edition of Bell's 'Quadrupeds,' however, the authors express their belief that "it will be found a characteristic of all ages, although certainly more marked in the young than in very old animals."

Dr. Brown says that the Greenland Seal (*Ph. grœnlandica*) in its second coat has often been mistaken for this species, but that the former may readily be distinguished by its having the second toe of the fore-flipper the longest. The hair next the skin is short and woolly, but externally harsh and shining, admirably adapted for repelling the water in which the animal passes so much of its time; the whiskers with which the upper lip is furnished, are thick, flattened hairs, laterally compressed, presenting diamond-shaped inequalities: this form of bristle is found in all the British Seals, whereas *Phoca barbata*, a species shortly to be mentioned as of doubtful occurrence on our coast, has the bristles compressed, but smooth. The food of the Common Seal consists of fish and crustacea.

THE RINGED, OR MARBLED SEAL.

The only recorded instance of the occurrence of the RINGED SEAL, *Phoca hispida*, of Schreber, on the British coast, is that of an individual captured on the Norfolk coast, in June, 1846, and purchased by Mr. J. H. Gurney, in the flesh, in the Norwich fish-market, the skull of which is now in the Museum of that city. Although no other instance of its occurrence is on record, it seems not improbable that it may occasionally be met with, and pass unrecognized. In the first volume of the 'Magazine of Zoology and Botany,' Mr. Wilson, in a paper on the Scottish Seals, speaks of a small Seal which was sometimes seen in the Hebrides, and believed by the natives to be a distinct species: this was rendered probable by their not associating with the Common Seals, and not being so wild in their nature. It is thought that this small Seal may have been *Ph. hispida*. Small dark-coloured Seals have more than once been seen on the Norfolk and Lincolnshire coast, or exhibited in the towns, which it is quite possible also may have belonged to this species. That it inhabited the coast of Scotland

Fig. 3. RINGED SEAL. (*Phoca hispida*).

in the past, there is evidence in the abundance of the remains of this species found in the glacial clays of that country, as identified by Professor Turner.*

The small Seal found in the inland fresh-waters of Lake Baikal is believed to be a variety of this species, differing only in its darker colour; it has, however, been separated, under the name of *Ph. baikalensis* by M. Dybowski (*Arch. f. Anat. u. Phys.*, 1873, p. 109). The type of *Ph. discolor*, F. Cuv., was taken in the Channel, and, according to De Sélys-Longchamps, this species has also occurred on the Belgian coast.

At present its home is the high latitudes of the Arctic seas, especially parallels 76 and 77 deg. North, and many are killed in South Greenland. In Davis's Straits it is found all the year round, particularly up the ice-fjords; in Cumberland Gulf it is said to be by far the most common Seal, and forms the principal food of the Esquimaux. This was the only species found by the late Arctic expedition north of Cape Union, 82° 15′ N. lat. Captain Feilden, the Naturalist to Sir G. Nares' Arctic Expedition, in an account of the 'Mammalia of North Greenland and Grinnell Land' (*Zoologist*, 1877, p. 359), thus speaks of this species:—" The Ringed Seal was met with in most of the bays we entered during our passage up and down Smith Sound. It was the only species seen north of Cape Union, and which penetrates into the Polar Sea. Lieutenant Aldrich, R.N., during his autumn sledging, in 1875, noticed a single example in a pool of water near Cape Joseph Henry, and a party which I accompanied in September, 1875, secured one in Dumbell Harbour, some miles north of the winter quarters of the "Alert": its stomach contained remains of crustaceans and annelids. In June of the following year I observed three or four of these animals on the ice of Dumbell Harbour. They had made holes in the bay ice that had formed in this protected inlet. The polar pack was at this

---

* *Journal of Anatomy and Physiology*, 1870, p. 260.

time of the year firmly wedged against the shores of Grinnell Land, and so tightly packed in Robeson Channel that no Seal could by any possibility have worked its way into this inlet from outside. I am, therefore, quite satisfied that *Phoca hispida* is resident throughout the year in the localities mentioned. A female killed on the 23rd August, 1876, weighed 65 lbs." This species has, therefore, probably the most northerly habitat of any existing mammal.

Dr. Brown, in his paper on the 'Greenland Seals' ('*Proc. Zool. Soc.*, June, 1868,) gives an interesting account of this species, which, like the preceding, is littoral in its habits, seldom frequenting the open sea, but found generally in the neighbourhood of the coast ice, in retired situations. It is known by the whalers as the "Floe rat," and its food consists of various species of crustacea and small fishes. This is the smallest of the Northern Seals, and of very little commercial value: its flesh, however, is eaten, and its skin forms the chief material of clothing in Greenland.

In appearance, this species is very like the Common Seal; but it is darker in colour, more particularly on the back, and the spots in the adult are surrounded by oval-shaped whitish rings; the young ones are lighter in colour. The old male is said to emit a most disgusting smell: hence one of its specific names, "fœtida." Dr. Rink says that this unpleasant odour is more developed in those which are captured in the interior ice-fjords, "which are also, on an average perhaps, twice as large as those generally occurring off the outer shores. When brought into the hut, and cut up on its floor, such a Seal emits a smell resembling something between that of assafœtida and onions, almost insupportable to strangers. This peculiarity is not noticeable in the younger specimens, or those of a smaller size, such as are generally caught, and at all events the smell does not detract from the utility of the flesh over the whole of Greenland." *

---

* 'Danish Greenland, its People and its Products,' p. 123.

Fig. 4. GREENLAND SEAL (*Phoca groenlandica*).
Adult and Immature.

The molar teeth in this species are arranged in a straight line along the jaws, and not obliquely, as in the common species. As this Seal is very likely to pass unnoticed, should it occur on our coast, it will be well to bear in mind that this arrangement of the molars will at once distinguish it from *Ph. vitulina*, the only species with which it is likely to be confounded. Professor Flower has given a minute description of the skull of the Norfolk specimen in the '*Proc. Zool. Soc.*' for 1871, pp. 506-12. The figure of this species is copied from Karl Thorin's 'Grundlinier Zoologiens Studium,' p. 53 (Stockholm, 1868).

THE GREENLAND SEAL.

The claims of the GREENLAND SEAL, *Phoca grœnlandica* (Fab.), to a place in the British Fauna, although long considered highly probable, were not rendered perfectly conclusive until 1874, when they were satisfactorily established by Professor Turner's identification of a Seal killed in January, 1868, near the viaduct on the Lancaster and Ulverstone Railway, and now preserved in the Kendal Museum. Professor Turner ('*Journal of Anatomy and Physiology*,' vol. ix. p. 163) says that he has himself examined this specimen, and found the dentition exactly to agree with that of the skulls of the Greenland Seals with which he compared it. The individual in question, a male, measured six feet from the tip of the nose to the "point of the hind toes," and the colour indicated the age to be about three years. Previously to this, the claims of this species to a place in our list rested principally upon the skulls of two Seals killed in the Severn, and exhibited by Dr. Reilly at the meeting of the British Association at Bristol in 1836. These skulls were at first referred by Professor Nilsson to *Ph.*

hispida, but afterwards, both by that gentleman and Professor Bell, determined to belong to *Ph. grœnlandica*. Doubts having been thrown on the accuracy of this decision, Professor Bell, in the second edition of his 'British Quadrupeds' p. 253, again states his belief that he was correct in assigning them to the young of this species. These specimens are unfortunately lost. Several supposed cases of the occurrence of this species are recorded, but in no instance were they supported by the production of the animal itself. Dr. Saxby ('*Zool.*' 1864) says that this Seal is not rare in bad weather in the Voe of Baltasound, Shetland ; and Mr. H. Evans, of Darnley Abbey, Derbyshire, in the year 1856, shot what he believes to have been a Greenland Seal near Roundstone, county Galway,—" Unfortunately, the animal sank and was lost; but Mr. Evans, who is well acquainted with the common and grey species, is perfectly certain that it was quite different from either" (Bell, 2 edit., p. 254). Perhaps the best authenticated case of the supposed occurrence of this species on our shores is given by Mr. H. D. Graham in Part I., vol. i. of the 'Proceedings of the Nat. Hist. Society of Glasgow,' p. 53 (Feb. 24, 1863). Three large white Seals were seen by Mr. Graham in Loch Tabert, Jura, Western Isles, lying on some shelving rocks, about 300 or 400 yards from the shore. They were watched through an excellent deer-stalking telescope for three hours, and Mr. Graham states that the characteristic markings of the Harp Seal could be distinctly seen. He also believes that, in three authentic instances, captures of *white* Seals, of extraordinary size, had been made, and states some particulars of the habits and appearance of these animals, as communicated to him by the islanders— to whom they appear to have been well known,—which render it highly probable that they belonged to this species. Mr. J. A. Harvie-Brown[*] also saw four Seals, which he believes to have been of this species, on a rock in

---

[*] 'Mammalia of the Outer Hebrides,' *Proc. Nat. Hist. Soc. Glasgow*, 1879, p. 95.

the Sound of Harris, on May 2nd, 1870. They took to the water, but as they "kept close in, and often rushed past within a few feet" of where he and his companion were standing, they had an excellent view of them, and "the large splashy-looking dark marks on either side of the back" were distinctly visible. Although essentially an Arctic species, this animal has a very wide geographical range, which, added to its migratory habits, renders it not at all improbable that individuals occasionally wander to our shores.

This species is a native of the Arctic Ocean, and ranges from the N.E. coast of America to the Kara Sea (where it was found by the Swedish Arctic Expedition in 1875), changing its quarters according to season.* It is this species which constitutes the chief object of pursuit in the northern Seal-fishery, and the season chosen for the attack is when they visit the ice for the purpose of producing their young ones. Dr. Brown says, "They take to the ice, to bring forth their young, generally between the middle of March and the middle of April, according to the state of the season, &c., the most common time being about the end of March. At this time they can be seen literally covering the frozen waste, with the aid of a telescope, from the 'crow's-nest,' at the main royal mast-head, and have on such occasions been calculated to number upwards of half a million of males and females."† The young, when born, are pure white, which changes to a yellow tint. At about 14 days old they begin to take to the water, and at the age of a month are capable of taking care of themselves: they then assume a spotted coat, which changes gradually to the adult markings, which are perfected in about three years. The adult male is about five feet long, the body generally of a tawny

---

* *Ph. grœnlandica* was the only Seal met with by the Austrian Arctic Expedition, in the *Tegethoff* n August, 1873, the ship then drifting in the ice in lat. 79° 31′, long. 61° 43′. Subsequently both this species and *Ph. barbata* were met with about North lat. 81°.

† 'Seals of Greenland.' Reprinted in '*Manual and Instructions for the Arctic Expedition*, 1875,' p. 47.

grey, varying to nearly white, marked with a conspicuous band of dark brown or black spots running into each other, which, commencing on the upper part of the back between the shoulders and curving downwards, is continued along the sides, disappearing before it reaches the hind flippers. The under parts are a dingy white, and the muzzle nearly black. The female, according to Dr. Brown, rarely reaches five feet in length, and is a dull white or yellowish straw-colour, tawny on the back, and with similar markings to the male, but somewhat lighter. Some are bluish or dark grey on the back, with "oval markings of a dark colour apparently impressed on a yellowish or reddish-brown ground:" these, Dr. Brown believes to be young females. The adult Greenland Seal is readily recognized, but it varies so greatly in its different stages of immaturity, and individuals differ so much from each other, that the most trustworthy characters are to be found in the dentition and the structure of the skull, which should in all cases be preserved, as affording the most ready and reliable means of determining the species of doubtful individuals. As has before been said, the second toe of the fore flipper is the longest in this species.

HOODED SEAL.

The HOODED OR BLADDER-NOSED SEAL, *Cystophora cristata* (Erxleben), fig. 5, has occurred at least thrice upon our shores. In June, 1847, a young one was killed in the Orwell, and is now in the Ipswich Museum; in 1872 a second young one was killed in Scotland near St. Andrew's; and a third specimen, an adult male, was caught in February, 1873, at Frodsham, on the Cheshire side of the Mersey, and lived in captivity till the beginning of the following June (Pr. Liverpool Soc. xxvii. p. 63). Others are believed to have

Fig. 5. Hooded Seal (*Cystophora cristata*).

been obtained in the Orkneys. Mr. Howard Saunders was assured that the "Bladder-nose" is well-known as a visitor to the Vae Skerries, Shetland (Alston's 'Mammalia of Scotland,' p. 15); and a Seal supposed to be of this species was seen off the Irish coast near Westport. In Hollingshed's 'Chronicles,' in the year 1577, sundry fishes of monstrous shape, with cowls on their heads like monks, and in the rest resembling the body of a man, are said to have occurred in the Firth of Forth (Bell's 'Brit. Quads.'), the appearance of which was of course followed by pestilence and famine. Throughout the Polar seas this species is widely distributed, being found in the Greenland seas, Iceland, and Spitzbergen, also occasionally in the temperate waters of Europe and America. It is polygamous and migratory in its habits: during the rutting season it is very pugnacious, and Dr. Brown says great battles take place between the males, and their roaring is said to be so loud that it can be heard for miles off. The young, which are born in April, are pure white at first, which changes to grey, and gradually becomes darker till it assumes the adult colour and markings, which it appears to do about the fourth year; the colour then is "dark chestnut or black, with a greater or less number of round or oval markings of a still deeper hue." The adult is furnished with a curious bladder-like appendage, commencing at the nostrils, with which it is connected, and continued upwards to the forehead : this, when inflated, presents a very remarkable appearance ; when the animal is at rest it remains flaccid, but when irritated or excited, it is blown up to its full extent. It is generally believed that the "bladder" is found only in the male, but Dr. Brown does not think there is any just ground for this belief; he does not, however, assign any reason for doubting what has been positively asserted to be the case. The Bladder-nose Seal is fierce in its nature and dangerous to attack ; although not actually taking the initiative it is always ready for battle, and will avail itself of any advantage by turning upon and following its opponent. The air-bladder, which is placed in the spot usually most vulnerable, renders it difficult to kill, as it forms a protection from the clubs of the sealers. This is one of the largest of the Northern Seals, varying, according to different authorities, from 7 to 10 or even 12 feet in length. The first toe of the fore flipper is the longest.

THE GREY SEAL.

One other species of true Seal, the GREY SEAL, *Halichœrus gryphus* (Fab.), claims a place in the British Fauna. Dr. Brown says the Grey Seal "has no doubt been frequently confounded with other species, particularly *Ph. barbata* and *Ph. grœnlandica.*" Such has undoubtedly been the case, and a specimen in the British Museum, long regarded as *Ph. barbata*, has been referred to this species. There is, I believe, no sufficient evidence that *Ph. barbata* has ever occurred on the British coast; but so imperfect even now is our acquaintance with the Seals which frequent our shores, that it may even yet be found. As before mentioned, the bristles forming the "whiskers' of *Ph. barbata*, are simple flattened hairs, without the impressed pattern found in the bristles of the known British species; they are nearly the same thickness throughout, and sharply curved near the end.

The Grey Seal has been found on various parts of the coast, from Shetland to the Isle of Wight; the Orkney and Shetland Isles, the Hebrides, and the west coast of Ireland, however, appear to be its chief places of resort on our shores; it has also been known to breed on the Fern Islands. Haskier Island, off North Uist, has long been known as a favourite breeding-place of this species. Captain Elwes, who visited this island on the 30th June, 1868 ('Ibis,' 1869, p. 25), informed Mr. Harvie-Brown that, up to the year 1858, an annual battue was held there in the month of November, when

Fig. 6. GREY SEAL (*Halichærus gryphus*).

the Seals resort to the rocks with their young ones, and that from forty to one hundred, old and young, would be killed. This wholesale destruction has been put a stop to, and as it is extremely shy and difficult to approach at other seasons, it is to be hoped that this species may for some time escape extermination in this favourite resort.

According to Bell, this species inhabits the "temperate northern seas rather than the Polar waters," and is found in the North Sea, Baltic, Iceland, Scandinavia, Denmark, and North Germany. Dr. Brown met with a specimen a little south of Discoe Island, but can only speak of its claims to a place in the Greenland Fauna as strongly probable. Bell gives some interesting information with regard to the habits of this species as observed in various British stations, and calls attention to the remarkable fact, that whereas in this country it produces its young in the months of October and November, on the Continent this is always said to take place in February; he suggests, to account for this singular discrepancy, that in our milder climate pairing takes place much earlier than in Scandinavia. The young, which are born white, are suckled for about a fortnight; the first coat is shed before they take to the water, which is not for some weeks after birth. The colour varies with age, sex, and season, so much, that it is not of great service in their identification, their large size being the best external guide. Lloyd, in his 'Game-birds and Wild-fowl of Sweden and Norway,' speaking of this species, says that even should it somewhat resemble the Common Seal in size and colour, as is at times the case, it may always be readily distinguished from the latter by the greater length of its claws and the superior breadth of its muzzle. The claws project considerably beyond the ends of the toes, the first of which is the longest. The general colour of the adult is greyish, tinged with yellow, and spotted and blotched with darker grey; the under parts lighter. The length of the adult varies from 7 to 10 feet. By the form of its skull and teeth it is readily distinguished, as well as by the great size of the animal. In the skull the brain-case is small, the nasal opening very large, and the grinders conical, only the two hinder pair in the upper, and the last pair in the lower jaw, double-rooted, the rest simple. Professor Bell, in his history of 'British Quadrupeds,' gives the generic and specific characters, as well as excellent figures of the skulls of the various British Seals, which will be found most useful in determining the species of any doubtful individuals; other figures will be found in Dr. Gray's 'Catalogue of the Seals and Whales in the British Museum.'

THE WALRUS, OR MORSE.

Of the many strange forms which the Zoological Society of London has been the means of introducing to the stay-at-home naturalists of this country, certainly not the least interesting is that of the Walrus (*Trichechus rosmarus*, Linn.) It is true that in neither of the instances in which the young animal has been brought alive to the Gardens, has it long survived in its new home; but, short though its residence amongst us, the opportunity has been afforded to many of becoming acquainted with the Arctic stranger in *propriâ personâ*, instead of through the distorted medium of the badly-stuffed skins, or the equally bad representations of this interesting animal, which, until recently, we have possessed. The first recorded appearance of the Walrus in this country was, I believe, in 1624, when, according to Hakluyt's 'Pilgrimes,' a young one was brought to England by Master Thomas Welden, in the *Godspeed*, and duly presented at Court. In 1853 the Zoological Society became possessed of a young one, which lived only a few days in their Gardens. On the 1st of November, 1867, another was received, which lived till the 19th of December, when it unfortunately died, notwithstanding the care

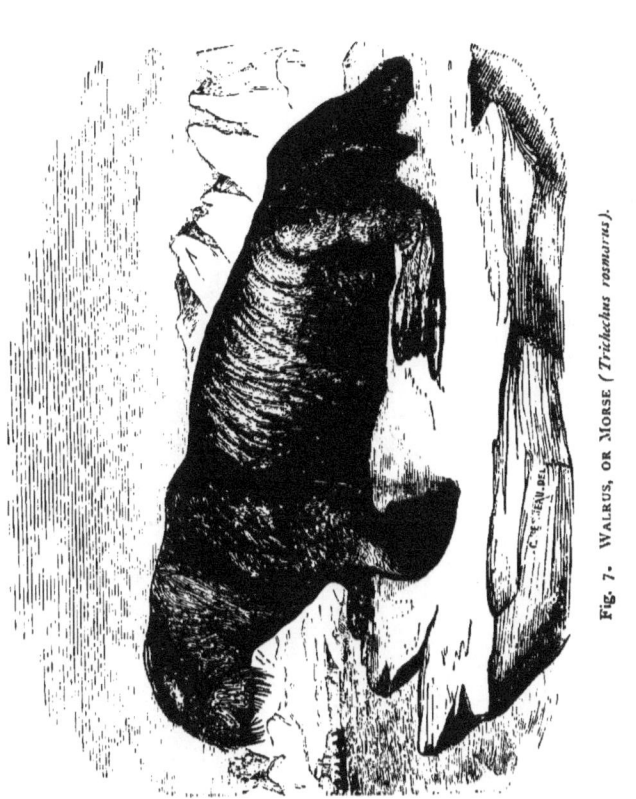

Fig. 7. WALRUS, OR MORSE (*Trichechus rosmarus*).

bestowed upon it, both as regards food and accommodation. This last was captured by the whale-ship *Arctic*, on the 28th of August, 1867, in lat. 69° N. and long. 64° W., and brought to Dundee, whence it was conveyed by Mr. Bartlett to the Society's Gardens. The captain of the *Arctic* saw two or three hundred walruses basking upon the ice, and sent out his boats to the attack: among the killed was an old female followed by her young one; the latter was taken on board and eventually brought to England.

Although now confined to the icy seas of the Arctic circle, the Walrus was probably not uncommon on our shores in times long past. The skull is said to have been found in the peat near Ely, and Hector Boece, in his 'Cronikles of Scotland,' mentions it as a regular inhabitant of our shores in the end of the 15th century: in the present century it has occurred several times, although it must be considered as a very rare straggler, sadly out of its latitude. Wallace says that its fossil remains have been found in Europe as far south as France, and in America probably as far south as Virginia, and it was common in the Gulf of St. Lawrence so late as 1770 (Leith Adams). In recent times it has retreated before its great enemy, man, from the northern coasts of Scandinavia to the circumpolar ice of Asia, America, and Europe, sometimes, but rarely, reaching as far south as lat. 60°. In Smith's Sound the Walrus does not appear to move further north than Cape Frazer, the meeting-place of the polar and southern tides: at this point Captain Feilden saw a single example. Whenever met with, it is the object of ruthless persecution, and is rapidly and surely becoming exterminated wherever man can reach it; and but for its ice-loving habits, which render its present strongholds always difficult, and sometimes impossible, of access, it would doubtless long ere this have become extinct.

Recently it has been met with on our shores, according to Bell, on the coast of Harris in 1817; in the Orkneys in 1825; one was seen in 1827 in Hoy Sound, but not captured; and in 1841 one was killed near Harris.

Dr. Brown also states that two were seen, one in Orkney and the other in Shetland, in 1857. Prof. Heddle also informed Mr. Harvie-Brown that in 1849 or 1850 he saw an adult, and a young one, off the coast of the parish of Walls, in Orkney (Harvie-Brown, *Proc. Nat. Hist. Soc. of Glasgow*, 1879, p. 97.)*

The *Trichechus* may be considered as intermediate between the true Seals and the Eared Seals, to both of which families it has affinities: it is carnivorous, feeding on mollusks, fish, and when it can get it, the flesh of whales. The stomach of one, examined by Captain Feilden, contained a large amount of green fluid oil, in which small particles of *Ulva latissima* could be detected, and minute fragments of the shells of *Mya*. Its habits were so well and succinctly described by Captain Cook a hundred years ago, that I cannot do better than quote his own words, the accuracy of which has since been amply confirmed. Whilst in Behring's Straits, in lat. 70° 6', and long. 196° 42', on the 19th of August, 1778, Cook first met with the Walrus: "they lie," he says, "in herds of many hundreds upon the ice, huddling one over the other like swine, and roar or bray very loud ; so that in the night, or in foggy weather, they gave us notice of the vicinity of the ice before we could see it. We never found the whole herd asleep, some being always on the watch. These, on the approach of the boat, would wake those next to them, and the alarm being thus gradually communicated, the whole herd would awake presently. But they were seldom in a hurry to get away till after they had

---

* A communication in *Land and Water* for Dec. 20, 1879, p. 524, signed "R. M.," states that about the 20th of June, 1879, a Walrus was seen off the west coast of Skye. "He was seen lying on a rock near the shore, on a fine calm evening, near enough to remove all doubt as to the identity of the animal. . . . The huge tusks were quite easily distinguished." On being disturbed, it is said to have rolled into the water, and swam a short distance to another rock, on which it was seen to climb; after a little time it again took to the water, and was seen no more. As no names are given, it is impossible to investigate this report, or to judge what degree of importance should be attached to it.

been once fired at, then they would tumble one over the other into the sea in the utmost confusion; and if we did not at the first discharge kill those we fired at, we generally lost them, though mortally wounded. They do not appear to us to be that dangerous animal some authors have described; not even when attacked. They are rather more so to appearance than in reality. Vast numbers of them would follow and come close up to the boats, but the

Fig. 8. *Vacca marina* (reduced from Gesner).

flash of a musquet in the pan, or even the bare pointing of one at them, would send them down in an instant. The female will defend the young one to the very last, and at the expense of her own life, whether in the water or upon the ice. Nor will the young one quit the dam, though she be dead; so that if you kill one you are sure of the other. The dam, when in the water, holds the young one between her fore-fins."* Since Cook's time the Walrus has learned to fear man, its only enemy except the Polar Bear, and is more difficult to approach. When wounded, or its young in danger, it has been known fiercely to attack the boats sent for its capture, striving to overturn them, and piercing their sides with its tusks: many serious accidents have been the result.

The number of Walruses killed annually by the Norwegian and Russian hunters is very considerable; probably nearly an equal number are wounded and lost. As the female produces only a single young one at a birth, which is said to remain with the mother nearly two years, "until its tusks are grown long enough to be used in grubbing up the shell mud at the sea-bottom," it will readily be imagined that the destruction is greatly in excess of the production, and that they are rapidly decreasing in numbers. A communication in the *Field* of March 27th, 1880 (p. 381), received from St. Francisco, points out even more serious consequences resulting from the reckless destruction of the Walrus than the mere extermination of a species, itself a matter of no small regret. "If," says the writer, "the whalers reach Behring Strait before the ice breaks up, they remain on the coast, and often hunt the Walrus for weeks together, with startling and serious results. Last year's campaign was considered successful, as about 11,000 Walruses were secured, most of them within the Arctic Sea. But to attain this result, *between thirty and forty thousand animals were killed*, so that only *one-third* of the number destroyed were actually utilised. There can be no doubt as to the ultimate consequence of such glaring imprudence; but last year they were so painfully apparent as to touch even the hearts of those who occasioned them. Not that the whalers were moved to compassion by the victims themselves, but by the sufferings of the human beings who were deprived of their chief souce of subsistence. The

---

* Cook's Last Voyage, vol. ii. p. 458, edition 1784.

hardy tribes in the neighbourhood of Behring Strait literally cannot exist without the Walrus, and so long as they were its only human enemies the number destroyed was inconsiderable. But the herds soon dwindled under the superior weapons and appliances of civilised nations, and the survivors retreated, like the Whales, towards the Pole. By the end of last season, not

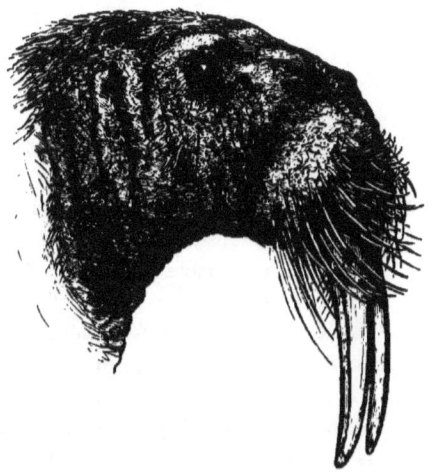

Fig. 9. HEAD OF WALRUS (Modified after Murie).

a single Walrus was left on the coast, and the immediate result was such a terrible famine among the natives that the whalers themselves speak of it remorsefully. The population north of St. Lawrence Bay has been reduced by one-third; and in a village which formerly contained 200 inhabitants, only one man survived. Several of the whalers have consequently refused to take any part in future Walrus hunts on the coast; they assert that for every hundred animals killed, a native family must perish by starvation, and they will not incur so heavy a responsibility."

About the month of August they repair to the shore, and congregating in vast herds on the beach of some secluded bay, lie for weeks together in a semi-torpid condition, without moving or feeding. Should their retreat be discovered whilst in this state, great is the slaughter. Mr. Lamont, in his 'Seasons with the Sea Horses,' says that in 1852, on a small island off Spitzbergen (one of the Thousand Islands), two small sloops discovered a herd of Walruses consisting of three or four thousand, nine hundred of which they succeeded in killing, only a small portion of the produce of which, however, they were able to carry away.

The colour of the Walrus is brown, paling with age, and the skin is thickly covered with short hairs; the adult reaches the length of 10 or 15 feet, or, according to some authorities, even more, and weighs from two to three thousand pounds. Its rounded head, heavy muzzle, thickly set with stout bristles, small, round blood-shot eyes, and formidable tusks, give to this animal a ferocious appearance which is foreign to its nature, except when greatly excited or at pairing time, when the old bulls are said to fight with great fierceness and determination. A full-grown Walrus will yield from five to six hundred pounds of blubber, the oil from which, however, is not so fine as that of the Seal. The ivory tusks were formerly much used by dentists; at present, I believe, owing to the introduction of vulcanite, very little is applied to that purpose. Mr. Lamont mentions 24 in. in length and 4 lb. each in weight, as the size of a good pair of bull's tusks: a pair in the Norwich Museum measure 32 in. in length, and the heavier of the two weighs 9 lb. 9 oz. The immensely elongated canine teeth which form the "tusks," are found in both sexes, but are shorter and more slender in the female than in the male. The skin of the Walrus is valuable for many purposes.

Few animals, so long known to man, have, when figured, been represented so inaccurately as the Walrus: the hind feet are almost invariably depicted

Fig. 10. "Sea Horse" (*After Cook*).

extended backwards, like those of the Seal (so also in stuffed specimens), whereas in the living animals they can be directed to the front, and serve as supports to the body in progression on the land or ice, in the same manner as the hind limbs of the eared seals. Dr. J. E. Gray, in an article 'On the

Attitudes and Figures of the Morse,' in the Proceedings of the Zoological Society of London for 1853, pp. 112-16, reproduces some of the wonderful prints of this animal from old authors, most of which are purely imaginary: fig. 8, p. 37, is copied from one of these. By far the best portrait known, till quite recently, is one published in Amsterdam in 1613, where an old female and her young one are very accurately depicted: this has been reproduced in Bell's 'British Quadrupeds,' 2nd edition, p. 269. Fig. 10 is copied from the "Sea Horse," in the foreground of Cook's illustration in 'A Voyage to the Pacific,' &c., 1784 edit., vol. ii., p. 446; as will be seen, this figure forms the source from which most subsequent illustrations were derived. Fig. 7 is taken, by kind permission of the late Mr. F. Buckland, from his 'Log-book of a Fisherman and Zoologist,' and represents "Jemmy," the young Walrus, whose brief sojourn in the Zoological Gardens has already been referred to. One of Mr. Wolf's "Zoological Sketches" represents a herd of Walruses in almost every conceivable attitude, and of course beautifully drawn and coloured.

Some authors recognise two distinct species of Walrus, one of which is said to be confined to the northern shores of the Atlantic, the other to the Pacific Ocean. Mr. Allen, in the 'North American Pinnipeds,' enters at length into the subject, and minutely describes the peculiarities which characterise each species. Reviving, after the example of Malmgren, the almost obsolete generic name of *Odobænus*, he describes the Atlantic Walrus under the name *O. rosmarus*; the animal found in the Pacific he calls *O. obesus*. The chief external points of difference in the latter appear to be in the facial outline, the longer and thinner tusks, "generally more convergent, with much greater inward curvature; the mystacial bristles shorter and smaller, and the muzzle relatively deeper and broader, in corelation with the greater breadth and depth of the skull anteriorly." The eyes are also said to lack the "fiery red" appearance attributed to the Atlantic Walrus, and to be smaller and very protuberant. Cook's figure reproduced at p. 41, also that at p. 177 of Scammon's book, are those of *Odobænus obesus*, and the fine pair of tusks mentioned at p. 40, as now in the Norwich Museum, were probably also obtained from a Pacific Walrus. The figure at p. 33, and the excellent figure by Wolf, at p. 457 of Lloyd's 'Game-birds and Wild-fowl of Sweden and Norway,' are of the Atlantic Walrus.

It is much to be regretted that the extinction of this harmless and useful animal is merely a matter of time, and that perhaps before many years have passed it may have ceased to exist; the only hope appears to be that when it has become too scarce to render its pursuit remunerative, a remnant may still be left to continue the species around the far-off and unapproachable islands of the Arctic seas. Even in Franz Josef Land, where, in the summer of 1880, Mr. Leigh-Smith found the Walrus very abundant: it will probably not long remain unmolested, for that gentleman informed Captain Feilden that the Norwegian walrus-hunters, when they heard of his discovery, talked of pushing on for Franz Josef Land next summer, the Spitsbergen walrus-hunting having become very uncertain, from the paucity and wariness of the animals.[*]

---

[*] 'Some remarks on the Nat. Hist. of Franz Josef Land,' by H. W. Feilden, F.G.S., &c.—a Paper read before the Norfolk and Norwich Naturalists' Society, Dec. 28, 1880.

CETACEA.

The occasional stranding upon our shores of some monster member of the order CETACEA serves from time to time to reawaken our interest in these wonderful animals, and sets us thinking how little we know about them, and how small is our acquaintance with their life-history.

Nor is this lack of information surprising when we consider that the difficulties in the way of studying the larger Cetacea, are so great as to be almost insuperable to any ordinary person, and even to the leaders of zoological science rarely does the opportunity present itself of examining specimens in the flesh ; for, of the rare instances in which they are cast ashore, the majority occur in wild and unfrequented parts of the coast, where they are probably cut up for their oil before a naturalist has an opportunity of examining them. From their unnatural position when cast up, and their altered appearance, owing to the falling in of some parts and the distension of others, correct portraiture is almost impossible ; and their great size renders it difficult and expensive to make them serviceable to science, while from the putrid condition in which they are frequently found, a close examination is too often anything but agreeable. If seen in their native element, where alone they *should* be seen duly to appreciate their grand proportions and perfect adaptation to their mode of life, the view must be brief and too often distant, certainly affording rare opportunities for close observation. There is thus little left for naturalists to study, except the bony skeletons, and of these often mere fragments. Under these circumstances, we shall cease to wonder at the great confusion which, till recently, existed in the classification and nomenclature of the *Cetacea*, and which has been only partially cleared away, chiefly by the labours of Professors Flower and Turner in this country, and by Professors Eschricht, Reinhardt, Van Beneden, Gervais, and others on the continent. The literature of the subject is widely scattered and difficult of access; and, although Dr. Gray and Professor Flower have done much to condense and systematize what is known, our acquaintance with the tropical and southern species of this interesting order is not at present sufficient to furnish materials for a monograph worthy of the subject. No class of animals has been called so many names, or so vilely caricatured in portraits, as the unfortunate Whales.

It is scarcely necessary now to say that the *Cetacea* hold a fully recognized place in the great class *Mammalia*, although this honour has not always been accorded to them. Ray classed them with the Fishes; and although Linnæus finally placed them in their true position, Pennant, following his earlier mistake, failed to do so. The members of this order, which includes the Whales proper, Narwhal, Dolphins, and Porpoises (with which, until recently, the Dugong and Manatees were improperly associated under the name of Herbivorous Cetaceans), bring forth their young alive. These are nourished by the female, which, for this purpose, is furnished with two inguinal mammæ. They are warm-blooded, and breathe by means of lungs, rendering frequent visits to the surface of the water necessary, as the animal can only respire when the orifice of the nostrils, called the blow-hole, which is placed on the top of the head, is above water. The breathing apparatus is very peculiar, being so modified that the air is admitted into the trachea without passing through the mouth; the Whale can thus breathe freely, provided the blow-hole be above water, even when its mouth is submerged or filled with water. There are no external ears, but a small aperture situated just behind the eye, communicates with a perfectly-constructed internal hearing apparatus, and this, as the water is an excellent conductor of sound, is all-sufficient. The food of the *Cetacea* consists of various forms of marine animals, from the Seal, which frequently forms a meal to the fierce Grampus, to the minute creatures which go to build up the giant form of the Right-Whale. Some possess numerous formidable teeth in both jaws; others have teeth in the lower jaw only; and in one section the teeth are only present in the embryo, but in their stead, from the upper jaw depend curious plates, arranged side by side, to which the name of *baleen* has been given. . The animal is encased in a layer of fat called "blubber," which lies beneath the skin, and serves to

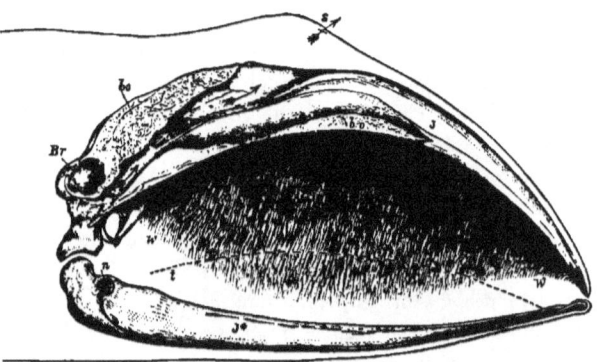

Fig. 11. MEDIAN SECTION, SHOWING OUTSIDE LEFT HALF OF SKULL. OF WHALEBONE WHALE, WITH BALEEN IN POSITION *(modified after Eschricht).*

Br., brain cavity; J, J*, upper and lower jaw-bones; bo, bo, being roughened parts of the bone sawn through; arrows indicate the narial passages, which open at s, spout-hole; w, whalebone; t, tongue, in dotted outline; n, nerve aperture, lower jaw.

retain the heat of the body, and the skin is smooth, polished, and quite devoid of hair or scales. On the back of most species is found a fleshy dorsal fin, and the fore limbs are represented by flippers externally undivided; the hind limbs, so far as external appearance is concerned, are altogether absent, but a rudimentary pelvis is found embedded in the flesh. The tail-fin forms the chief organ of locomotion: it is always fixed horizontally, and is of great size and power, enabling the animal, by its vigorous use, to attain great speed. There are many and striking peculiarities in the bony skeleton which it is not necessary here to enumerate.

Before proceeding to give some account of the species which have been found in the British Seas, it will first be necessary to say a few words as to the arrangement of the genera and species. I shall enter into this part of the subject, however, so far only as is necessary for us clearly to understand the relative positions of the species which we shall have to consider.

Professor Flower divides the order *Cetacea* into two sub-orders: First, *Mystacoceti*, or *Balænoidea*, in all the members of which baleen takes the place of teeth, which are never developed, disappearing before birth; second, *Odontoceti* or *Delphinoidea*, in which teeth (sometimes very numerous) are always developed after birth. The first sub-order is a very restricted one, embracing only two families, *Balænidæ* and *Balænopteridæ*, to the former of which belong the two genera of Right-Whales, *Balæna* and *Eubalæna;* and to the latter, two genera, namely, *Megaptera* and *Balænoptera*. To these two genera* belong the Rorquals, which occasionally occur in the British seas. The second sub-order, *Odontoceti*, contains the families of *Physeteridæ*, represented by the Sperm Whale, Beaked Whale, and several allied species; *Platanistidæ*, consisting of some curious forms found only in India and South America; and *Delphinidæ*, comprising the Narwhal, Beluga, or White Whale, Grampus, Porpoise, and Dolphins. The total number of British *Cetacea* has been variously estimated; Dr. Gray, in 1864, described thirty, and in 1873 thirty-three species; while Bell, whom we shall follow, recognised only twenty-two species in his second edition, published in 1874.

The following table of the British Cetacea will serve to indicate at a glance the precise position assigned to each species, in the two main divisions into which the order is divided:—

---

* *Physalus, Benedenia,* and *Sibbaldius,* of Gray, are now rejected, I believe, by Prof. Flower.

BRITISH CETACEA.

SUB-ORDER.	FAMILY.	SUB-FAMILY.	GENERA.	SPECIES.
1. MYSTACOCETI (Whalebone Whales.)	Balænidæ	Balæninæ	Balæna	(?) B. mysticetus, *Right-Whale* B. biscayensis, *Atlantic Right-Whale*
	Balænopteridæ	Megapterinæ	Megaptera	M. longimana, *Hump-backed Whale*
		Balænopterinæ	Balænoptera	B. musculus, *Common Rorqual* B. sibbaldii, *Sibbald's* ,, B. laticeps, *Rudolphi's* ,, B. rostrata, *Lesser* ,,
2. ODONTOCETI (Toothed Whales.)	Physeteridæ	Physeterinæ	Physeter	P. macrocephalus, *Sperm Whale*
			Hyperoodon	H. rostratus, *Beaked Whale* H. latifrons, *Broad-fronted Beaked Whale*
		Ziphiinæ	Ziphius	Z. cavirostris, *Cuvier's Whale*
			Mesoplodon	M. bidens, *Sowerby's Whale*
	Delphinidæ		Monodon	M. monoceros, *Narwhal*
		Beluginæ	Delphinapterus	D. leucas, *White Whale, or Beluga*
			Orca	O. gladiator, *Grampus, or Killer*
			Grampus	G. griseus, *Risso's Grampus*
			Globicephalus	G. melas, *Pilot Whale*
		Delphininæ	Phocæna	P. communis, *Porpoise*
			Delphinus	D. delphis, *Common Dolphin* D. tursio, *Bottle-nosed Dolphin* D. acutus, *White-sided Dolphin* D. albirostris, *White-beaked Dolphin*

MYSTACOCETI (WHALEBONE WHALES.)

BALÆNIDÆ.

THE GREENLAND RIGHT-WHALE.

The first species, both in order and importance, of the Family *Balænidæ* is the well-known *Balæna mysticetus*, the GREENLAND, or RIGHT-WHALE as it is called by the whalers. So extremely doubtful, however, are the claims of this animal to a place in the British Fauna, that it is retained in the present treatise solely on account of the great interest attaching to it as a species, and not from any idea of maintaining for it a position, which, although hitherto assigned to it, has now become untenable. The use of the term well-known is perhaps unadvised; for, although this species has engaged the energies and industry of the merchant seamen of Northern Europe for centuries, so little was known of it scientifically, that not a single skeleton had ever found its way into any European museum, until Eschricht obtained one from Holsteinborg, in Greenland, in 1846. The recorded instances of the supposed occurrence of this species in the British Seas are unsatisfactory in the extreme. The most positive record is that in Messrs. Paget's 'Natural History of Great Yarmouth.' They say: "*Balæna mysticetus*—common Whale—a small one taken near Yarmouth, July 8, 1784." Sir James Paget, however, in a letter to the Author, is unable to add to the brief statement, as will be seen from the following extract from his communication:—"I am sorry I can give you no information respecting the

Whale taken off Yarmouth in 1784; I have no notes as to the source from which I derived the statement, but probably it was from some MS. of Mr. Dawson Turner's. It is not likely that any bones of the Whale were kept in Yarmouth, for there was no naturalist there at the time, and the whaling-trade, which was then actively carried on from the port, must have made Whales' bones very common." This is all that is ever likely to be learned of the Yarmouth Right-whale; but the season at which it occurred would render the heated seas on our coast utterly unbearable to an ice-loving inhabitant of the Arctic seas. This, with its small size, would seem to point to a closely-allied species to be mentioned soon. Sibbald records the occurrence of what he considers was probably a Right-whale, at Peterhead, in 1682; and a Whale recorded at Tynemouth by Willughby may have been of this species. In the first edition of Bell's 'Quadrupeds' is a communication from the Rev. Mr. Barclay to the effect that on the coast of Zetland dead or very lean Whales of this species have several times been found or have run aground; but in the second edition of the same work the authors state that "there is no proof these references do not apply to some other species." The same may be said with reference to Low's remarks in the 'Fauna Orcadensis,' p. 158. This is all we know of the supposed occurrence of Right-Whales in British waters in recent times, and there is little doubt that these, if Right-Whales at all, should be referred to the next species.

The extreme northern habitat assigned to this species by those who have devoted much time and labour to the investigation of the subject, clearly proves that it must either have changed its habitat, which its present habits seem to render improbable, or that some other species formerly inhabited the temperate seas outside the Arctic circle extending southward to the Atlantic as far as latitude 40°, for it is beyond doubt that a brisk whale-fishery was carried on in former times by the Basque population in the Bay of Biscay and adjacent seas as far back as the 8th or 10th century. That such a

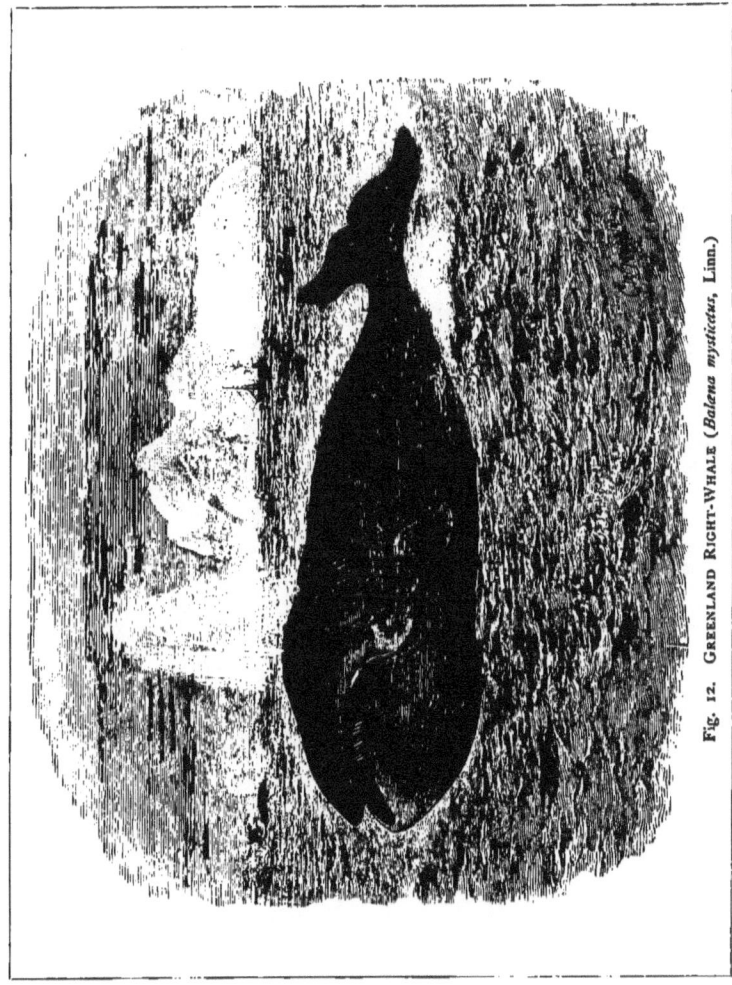

Fig. 12. Greenland Right-Whale (*Balæna mysticetus*, Linn.)

southern species, distinct from the northern Right-whale did exist, is proved by Professors Eschricht and Reinhardt in their splendid memoir of the 'Greenland Whale,' a translation of which, edited by Professor Flower, was published by the 'Ray Society' in 1866, and of that species we shall give some account further on.

It has been asserted that the Greenland Whales supposed formerly to have visited our coasts, have been driven north by the increased traffic in the more frequented seas of temperate Europe; but from the habits of this species as observed on the west coast of Greenland, at the fishing stations established by the Danish Government, and recorded in the memoir just referred to, no confirmation of this theory is afforded. The fishery at these stations was prosecuted from the shore when the Whales appeared upon the coast in the winter months; as the spring advanced they followed the receding ice-line, and were seen in summer as far north in Baffin's Bay as ships had at that time succeeded in penetrating, whilst their southward range in winter was always limited by a rather northerly degree of latitude. This, it is shown, went on with the greatest regularity for at least 80 years, during which the Whales constantly made their appearance at the same places, at the same season, without the slightest alteration taking place. The fact of the Whales always moving northward as the ice breaks up, will account for their being found in the spring in different latitudes; thus, on the Greenland coast, they are found, at this season, in latitude 65° 25'; but in Davis' Strait, in 61° to 62°, always, however, inseparable from the ice. Messrs. Eschricht and Reinhardt thus conclude: "It seems, therefore, that the Whales have not retreated further north, as they are still found within precisely the same limits in which they were found at the beginning of the persecution, but in numbers so diminished that the fishery will hardly repay the trouble and expense attending it."

Capt. Feilden, the naturalist to Sir Geo. Nares's Arctic expedition, speaking of the Northern range of this species, says he is quite satisfied that "no Whale could inhabit at the present day the frozen sea to the north of Robeson Channel. To penetrate from the North-water of Baffin Bay to Robeson Channel, would be a hazardous task for this great animal, and in this opinion the experienced whaling quartermasters, who accompanied our Expedition, coincided. We may dismiss from our minds the idea or hope that nearer to the Pole, and beyond the limits of present discovery, there may be haunts in the Polar Sea suitable for the Right-whale. I do not look for the speedy extinction of the Greenland Whale; but it is probable that in a few years the fishing will no longer prove profitable to the fine fleet of whalers that now sail from our northern ports, and I see no hope of Arctic discovery increasing our knowledge of the range of this animal."[*]

The southern limit of the Right-whale in the Northern ocean may be shown by a line drawn from the coast of Lapland at 70°, just touching the southern point of Iceland, and ending on the coast of Labrador at about 55° north latitude.

The whaling-trade, which once employed so many hardy seamen, is now reduced to very narrow limits, and appears to have passed almost entirely into the hands of the English, or rather Scotch. The Biscayans were not content with exterminating the Whales found in their own seas, but in 1721 they had twenty vessels in the Greenland fishery; the Dutch also took a large part in the trade; and in the year 1680, when they appear to have been the most actively engaged in the fishery, they are said to have had about 260 ships and 14,000 men employed. In 1725 the South Sea Company embarked in the trade, but meeting with considerable losses, speedily gave it up. The Government, in order to encourage this languishing branch of industry, in 1732 granted a bounty of 20s. per ton on the oil; this, being found insufficient,

---

[*] *Zoologist*, 1877, p. 360.

was increased in 1749 to 40s. per ton, which caused a considerable increase in the number of vessels; but upon Parliament, in 1777, reducing the bounty to 30s. per ton, the number of vessels rapidly fell off from 105 to 39; the bounty was then, in 1781, raised to its old level, with a corresponding increase in the number of vessels employed. Then followed a gradual process of reduction, until in the year 1824 the bounty altogether ceased, and the ships fell off from 112 in 1824, to 88 in 1827.* During the nine years ending 1818 there was an average of 91 English (sailing from eight ports), and forty-one Scotch ships (sailing from nine ports) employed in the trade; in 1830 they were reduced to 41 English vessels (sailing from five ports), to which Hull contributed 33, and 50 Scotch vessels (sailing from seven ports), to which Peterhead contributed 13, and Dundee 9.

The years 1819 and 1830 were both very disastrous to the whale-trade; in the former year fourteen British vessels were lost, and in the latter, nineteen British ships were totally wrecked, and twelve seriously injured. The number of ships employed has since gradually decreased, and at present Dundee and Peterhead are the only two ports in Great Britain engaged in the whale-fishery. Dundee sends out fifteen powerful steam vessels, which leave about the beginning of May, and if fortunate in filling up, return, according to circumstances, from August to the beginning of November. Peterhead sends five steamers and one sailing vessel; they are ship-rigged, and from two to five hundred tons register, and 40 to 100 horse power. The expense now incurred renders it necessary that a large number of Whales should be taken to make the voyage pay: the *Arctic*, in her voyage of 1873, captured twenty-eight Whales, which were estimated to produce in oil and bone £18,925, or about £678 per Whale, the best Whale, a female with sucker, was estimated at £1,500, and the smallest at only £110. An average Whale produces 9½ tons

---

* McCulloch's *Dictionary of Commerce*.

of oil, a ton measuring 252 gallons, and 7 ft. 6 in. of whalebone; the longest bone cut of the twenty-eight fish was 11 ft. 9 in. and the shortest 2 ft. 6 in. This was considered a very successful year. The whale-fishery was commenced at Peterhead in 1788; since that time, up to the year 1879, Captain David Gray informs me that 995 voyages have been made to the Greenland and Davis' Straits whale and seal-fisheries, and there have been brought home 4195 Whales, furnishing 30,975 tons of oil, and 1549 tons of whalebone, besides 1,673,052 Seals, yielding 20,913 tons of oil, leaving a nett profit of £583,020, or £586 per ship per voyage. The Dundee whale-fishery commenced in 1790, and the seal-fishery in 1860; since that time up to the season of 1879, 538 voyages have been made to the Greenland and Davis' Straits whale and seal-fisheries, including Labrador, which have produced 4220 Whales, yielding 32,774 tons of oil and 1640 tons of whalebone, besides 917,278 Seals, yielding 10,464 tons of oil, valued together at £2,160,400, leaving a nett profit of £652,320, or £1212 10s. per ship per voyage. Capt. Gray adds: "I have often been asked where all the Whales are gone to; let the above figures be the reply."

The present price of whale-oil is from £28 to £30 per ton, the whalebone ranging as high as £1100 per ton, according to the length of the bone; but although there are exceptions, of late years the fishery, as a whole, is said, on good authority, not to have paid the heavy expenses of the fleet engaged in it, nor does there seem much prospect of improvement, mineral oil being now used for many purposes for which formerly whale and seal oil was required. One of the chief uses to which whale and seal oil are now applied is in the preparation of the jute fibre, the manufacture of which is so extensively carried on at the port of Dundee, also the chief centre of the whaling trade.

An interesting account of a whaling voyage in the ship *Arctic*, and full particulars of the mode pursued in taking, and subsequent treatment of the fish, is given by Captain A. H. Markham, in his 'Whaling Cruise to Baffin's Bay.' *

The usual length of a full-grown Right-whale is about 50 feet; but Dr. Brown, in his paper on the Cetaceans of the Greenland Seas (*P. Z. S.*, 1868, p. 539), gives the dimensions of one which measured 65 feet. The general colour is black. The mouth occupies about one-third of the entire length, and the baleen is from 10 to 12 feet long; it has been known to reach the great length of 13 ft. 2 in., and 9 in. in width. This baleen, which is found depending from the upper jaw, consists of a number of horny plates, similar in structure to the horn of the rhinoceros, consisting of a fibrous mass glutinated together in the solid portion, and placed transversely along either side of the palate; they are arranged closely together, with the external edge smooth, and gradually thinning off towards the inner margin, which ends in a fringe of long hair-like fibres: the number of laminæ is about 300 on each side.† Captain David Gray, of the *Eclipse*, an experienced whaler, in a communication to 'Land and Water,' on December 1, 1877, pointed out and first satisfactorily explained the means by which these extraordinary appendages are disposed of when the mouth of the Whale is closed. He shows

---

* Space will not permit of more than a passing reference here, but much information as to the rise and progress of the whale-fishery will be found in McCulloch's 'Dictionary of Commerce,' article "Whale-fishery;" Scammon's 'Marine Mammals of the North-western coast of North America;' Starbuck's 'History of the American Whale Fishery;' Mr. C. R. Markham's 'The Threshold of the Unknown Region;' Capt. A. H. Markham's book above referred to; and above all in Scoresby's excellent works, which have been extensively laid under contribution by nearly all subsequent writers— 'An Account of the Arctic Regions, with a History and Description of the Northern Whale-fishery' (2 vols., 1820), and 'A Journal of a Voyage to the Northern Whale-fishery,' in 1822.

† Blackstone mentions a curious old feudal law, to the effect "that on the taking of a Whale on the coasts, which is a royal fish, it shall be divided between the king and queen; the head only being the king's property, and the tail of it the queen's. '*De Sturgione observetur, quod rex illum habebit integrum: de balena vero sufficit, si rex habeat caput, et regina caudam.*' The reason of this whimsical division, as assigned by our ancient records, was, to furnish the Queen's wardrobe with whalebone"!— Blackstone's 'Commentaries,' 1783 edit., vol. i., p. 223.

that when the mouth is shut, the slender ends of the whalebone curve backwards towards the throat, the longer ones from the middle of the jaw falling into the hollow formed by the shortness of those behind them; when the animal opens its mouth to feed, the whalebone springs forward and downwards, thus always by its elasticity, filling up the space between the upper and lower jaws, whether the mouth be fully or only partially open, and interposing a strainer between the cavity of the mouth and the external water, effectually preventing the food which enters the mouth from passing out with the flow of water which passes through the mouth as the great beast pursues and captures its minute food.

The Whale whilst feeding swims along with its mouth open, until it has collected a quantity of the small marine animals which form its food; then, closing its capacious under-jaw, it forces out the water between the plates of baleen, leaving the captive prey stranded on its huge tongue, when it swallows them at leisure. The food of the Greenland Whale consists entirely of small marine animals, particularly a kind of shrimp, found in great abundance in the Arctic seas. This species seldom remains under water longer than from ten to fifteen minutes, returning to the surface to breathe, which, if undisturbed, occupies from two to three minutes. Capt. Gray, however, has known it when harpooned to stay under water fifty minutes. Professor Owen describes the wonderful provision for storing of blood in a vast plexus of blood-vessels found in the Cetacea, at the back of the lungs and between them and the ribs, thus enabling them, although lung-breathing animals, to stay under water for so protracted a period, and states that the peculiar non-valvular structure of the veins of the Cetacea, and the pressure on these reservoirs of blood at the depths to which they retreat when harpooned, explain the profuse and lethal hæmorrhage which follows a wound, that in other mammalia would not be fatal.*

---

* Owen, 'Anat. of Vert., iii., pp. 546 and 553.

Fig. 13. ATLANTIC RIGHT-WHALE (*Balæna biscayensis*, Eschricht), after Capellini.

The Right-Whale is believed by Eschricht and Reinhardt to bring forth its single young one (rarely two) about the end of March or beginning of May, and the time of gestation to be thirteen or fourteen months, so that it will bring forth only every other year; Scoresby considers that they go eight or nine months, and bring forth in February or March.* The young one is supposed to be suckled for twelve months, during which time the baleen is gradually developed. In disposition, the Greenland Whale is timid and retiring; the chief danger in its capture arises from its rapid descent when harpooned; the line is then carried out with such speed that, should it foul or all run out and not be immediately cut, the boat will be upset or carried under water. Capt. David Gray estimates the speed of a struck or scared Whale at about eight miles an hour, and the ordinary speed at about four miles, whether sounding or along the surface. It has never been known to attack a boat, but accidents sometimes happen if approached too closely in its death "flurry," which is said to be very terrible to witness. Its fondness for its young is such that if the "sucker" be killed the old one readily falls a victim, and the whalers do not fail to avail themselves, for their own advantage, of this amiable trait in its character.

THE ATLANTIC RIGHT-WHALE.

Until recently it was believed that a Whale formerly common in the temperate waters of the North Atlantic was identical with the Right-Whale of the Arctic seas, of which we have just given an account, but Professors Eschricht and Reinhardt have successfully shown, as stated in the previous

---

* Dr. Brown, in the paper before quoted, states that they couple from June to August, and bring forth in March or April. See also a note on 'The Time and Manner of the Procreation of some Species of Whales,' in the 'Zoologist' for 1845, p. 1161.

article, that such is not the case, the habits of the two animals, as well as the localities frequented by each, being totally distinct. They have, therefore, described the more southern form as a distinct species, under the name of *Balæna biscayensis*, or the ATLANTIC RIGHT-WHALE, the "Sarde" of the French, "Nordkaper" of the Dutch, and "Sletbag" of the Iceland whalers of former days.

As early as the twelfth century, long before the whale-fishery was prosecuted in the Arctic seas, a brisk trade was carried on by the Basque fishermen from the Biscayan ports. That this fishery must have been of considerable importance, in a mercantile point of view, there can be no doubt, from the numerous references to be met with in early records; for instance, in 1261, a tithe was laid upon the tongues of all Whales imported into Bayonne, where they formed a much-esteemed article of food, and in 1338 a duty of £6 a Whale on those brought into the port of Biarritz was relinquished by Edward III. to Peter de Puyanne for services rendered; these and other like records extant show that for a long period this branch of industry was briskly prosecuted. Gradually, however, the Whales became more and more scarce, and the hardy Basque seamen, after following their prey to Newfoundland and Iceland, shortly after the discovery of Spitsbergen in 1596 found their all-but-lost occupation suddenly revive; the "Sletbag" was left behind, but the home of the true Greenland Whale, a much more valuable animal, was for the first time invaded, and that species, which then abounded in the seas surrounding Spitzbergen, speedily became the object of the whalers' attack; many vessels were fitted out for its pursuit which carried Biscayan harpooners, the crews, also, generally consisting, in part, of these hardy seamen.

So recently as the close of the last century, the Atlantic Right-whale was not unfrequent in the North Atlantic; it was regularly caught on the coast of Nantuckit, and occasionally by the American Whalers on the coast of Iceland; it has, however, now become very rare. Professors Eschricht and

Reinhardt thus summarise the distinctive characters of the "Sletbag," "Sarde," or "Nordkaper," so far as they have been able to glean from all the sources accessible to them, and consider the species identical with their *B. biscayensis*:—

1. "That it was much more active than the Greenland Whale, much quicker, and more violent in its movements, and, accordingly, both more difficult and more dangerous to catch."
2. "That it was smaller (it being, however, impossible to give an exact statement of its length), and had much less blubber."
3. "That its head was shorter, and that its whalebone was, comparatively speaking, much thicker, but scarcely more than half as long as that of the Greenland Whale, being, however, still much longer than that of even the very largest Fin-Whale, although the 'Sletbag' itself probably scarcely attained to half the length of the last named."
4. "That it was regularly infested with a Cirriped belonging to the genus *Coronula*, and that it belonged to the temperate Northern Atlantic as exclusively as the Greenland Whale belonged to the icy Polar Sea, so that it must be considered as equally exceptional when either of these species strayed into the range of the other, and, moreover, that in its native sea it was to be found farthest towards south in the winter (namely, in the Bay of Biscay, and near the coast of North America, down to Cape Cod), while in the summer it roved about in the sea round Iceland and between this Island and the most northerly part of Norway."[*]

In addition to the British Right-Whales mentioned at the commencement of the previous article, which may almost with certainty be referred to this species, I am enabled, through the kindness of my friend, Captain David Gray, of Peterhead, to record two other instances of the occurrence of the Atlantic Right-Whale in British waters. With regard to the first case, Captain Gray was good enough to obtain for me the independent testimony of two old men, James Webster and John Allan, both of whom are still living at Peterhead, and were witnesses of the events which they relate. The two

---

[*] 'Recent Memoirs on the Cetacea, by Professors Eschricht, Reinhardt and Lilljeborg,' edited by Prof. Flower, Ray Society, 1866.

statements coincide so remarkably, making allowance for the lapse of so many years, that it is only necessary to give one. "James Webster, 85 years of age, remembers Greenland Whales coming into South Bay of Peterhead : at that time he would have been about 10 years of age [Jno. Allan says "it was in 1806 .or 1807, same year as the new parish church was opened ; " this was in 1806, and agrees with Webster's statement that he was 10 years old at the time]. Remembers them being an old Whale and a sucker. Saw five boats go out after them; as far as he recollects, thinks it was the month of October ["in the summer-time," Allan]. They struck the old Whale, and put three harpoons into her, then they struck the sucker and killed it; brought the sucker ashore and flenched it at the South Quay. [Allan says "they killed the young Whale, and flenched her at the South Quay: she, having sunk, it was two or three days after, before they got her in." After they had three harpoons in the old Whale, she went twice up into the head of the Bay, going so far that she turned the sand up, and then she stove two of the boats, and broke Mackie's, one of the harpooners, legs. [Allan does not remember the name of the injured man, and thinks only one boat was stove.] After this, the Whale took a run, and went out of the Bay, blowing blood. They followed her as fast as they could, they cut two of the boats from her, and left her towing one boat with their Jack blowing, after taking the crew out of her, and in this condition the Whale went out of sight, and they never saw or heard of her again. Allan says that when she went round the South Head, a heavy sea being on at the time, and darkness coming on, the boats cut and let her go, leaving the boat which was stove, fast to the Whale, the flag still blowing, and that she went out to sea and was never seen again. Capt. Gray adds that " Capt. Wm. Volum, of the ' Enterprise,' and Capt. Alex. Geary, of the ' Hope,' both took part in the chase, and in that year the 'Hope' returned from Greenland on 30th June, and the ' Enterprise' on 30th July ; consequently, it must have been some time after the latter date that the Whales came into the Bay; probably Webster is right when he names October."

The second instance referred to by Captain Gray came under his own observation. Whilst taking a walk round the "Heads," one Sunday morning before church, to the best of his recollection early in October, 1872, "I saw," says Captain Gray, "a Greenland Whale within half a mile of the rocks off the South Head; its appearance and movements were exactly the same as those I have seen in Spitsbergen waters." Accustomed, as Captain Gray has been for many years, to watch the appearance and actions of the northern species of Right-Whale, in the Polar seas, it seems impossible for a man of his great experience to have mistaken any other species of Whale for one of the *Balæninæ*.

Of course, there still remains the question as to whether these Whales were the Greenland or Atlantic species, but I think the consideration of the circumstances under which they occurred, leaves no doubt what the reply must be. Captain Gray writes—"Until you began to question the identity of these Whales harpooned here in 1806, no one had ever had the smallest doubt of their being Greenland Whales," and that had there been any marked difference in their appearance, it would have been at once noticed by such experienced men as those who captured the Whale at Peterhead; but he adds that "so far as the habits of the Greenland Whale are known, it is contrary to our experience that they should visit our shores at the season of the year at which these Whales were seen here, when we know that the Arctic Whale regularly disappears into the depths of the Polar ice towards the end of summer, where no ships or steamers can follow them." It would naturally be expected that, towards the end of summer the Atlantic Whale would also be approaching the northern limit of its range, and this is precisely the season when all the Whales of this description, of which the date is given, appear to have occurred, except two in a much more southerly locality, (their proper winter habitat) shortly to be mentioned. That the Peterhead men did not speak of any marked difference in the Whale which visited their Bay and those they had just returned from pursuing in the Polar ice may perhaps be accounted for partly by the similarity of the two species, and partly by their not having killed the adult individual; whilst the restless activity of the latter may possibly be due, not only to the presence of its young one, but, in part, to the superior activity of the Atlantic species, which is said to render it so much more dangerous and difficult to catch.

But it may be said that if there be such a species, having a range, which in summer extends from the entrance of Davis' Strait to Iceland and the North Cape, why are they not occasionally met with by the whalers in crossing the Atlantic to and from their more northern fishing grounds? Although such an encounter with a creature confessedly of rare occurrence would be in the highest degree improbable, still here again, through the kindness of Capt. Gray I am able to say that such encounters have taken place, and could we know the experience of all the whalers who have crossed the Atlantic, perhaps other instances might be put on record. Captain David Gray's father told him that while mate to his father (Capt. David Gray's grandfather), when crossing the Atlantic on the homeward voyage from Davis' Strait, the vessel ran into a Greenland Whale (as he supposed it) and that he was anxious to lower some boats and go after it, but that his father would not allow him to do so, there being too much sea running at that time. This again would be in the summer season. It seems probable that not being aware of the existence of a Southern species of Right-Whale, or in consequence of the high sea which was running at the time, the Grays did not observe, or, at least, failed to mention, the peculiarities which distinguish the Atlantic species. But I am indebted to Capt. Gray for other instances of the occurrence of this species not far from Cape Farewell, and in at least one case the species was identified, the observer being aware of the existence of the Atlantic Whale, and the circumstances apparently favourable for close observation. On the 1st May, 1868, Capt. Alexander Murray, now commanding the S.S. "Windward," at that time trading to South Greenland, in the "Sir Colin Campbell," saw near Cape Farewell, several Right-Whales, close enough to distinguish their different features and general appearance. Capt. Murray remarks that, "they are a shorter Whale than the Greenland and much flatter in the crown;" he also noticed "Barnacles and grass near the blow-holes," and states that from conversations he has had with American shipmasters employed in hunting these Whales, that these parasites are always present in this species, whereas the Greenland Whales are as invariably free from them. Capt. Murray adds that in 1867 three American whalers came into Cumberland Gulf, one having six, one three, and the other two Atlantic Whales on board, all of which were taken in the summer, a little to the eastward of Cape Farewell; and, finally, Capt. Gray's brother, who commands the Hudson Bay Company's Steamer, "Labrador," told him that in June, 1879, he saw two of these Whales in lat. 57 N. and long. 33 W.; they were close alongside, and the weather at the time calm: they went away in a south-westerly direction. It would seem, indeed, that this species is not at all an infrequent summer visitor to the open sea, lying to the east of Cape Farewell.

Two recent instances of the occurrence of this species on the eastern side of the Atlantic are on record, both of which were met with in winter, and in the warmer latitudes of the Bay of Biscay and the Mediterranean Sea. On the 17th of January, 1854, a young one with its mother appeared in the harbour of St. Sebastian; the mother escaped, but the little one was caught, and a drawing of it made by Dr. Monedero (reproduced in Bell's 'Brit. Quad.,' 2nd Edit. p. 387); the skeleton was preserved for the museum of Pampeluna, thence it was removed by Prof. Eschricht in 1858 to the Copenhagen Museum, for which he purchased it. Also, on the 9th February,

1877, a Whale was captured in the Gulf of Taranto, which has been referred to this species, and these, I believe, are the only specimens which have been taken in European waters of late years; it seems very probable, however, that the "Black-Whale" of the temperate shores of N. America (the *B. cisarctica* of Cope) is identical with *B. biscayensis*, and that, although extinct on the eastern side of the Atlantic, individuals from the American waters occasionally find their way into the European seas, where the race formerly existed as a native. The skeleton of the Taranto specimen is now in the Museum of Comparative Anatomy of the University of Naples, and M. F. Gasco states positively that "both the Taranto Whale and that of Philadelphia (*B. cisarctica*, Cope) belong to the species *B. biscayensis*, of Eschricht, which, for several centuries was pursued with avidity—I was going to say exterminated—throughout the temperate regions of the North Atlantic, first by the Basques, and then successively by the Saintongeois, the Normans, the Dutch (who called it *Nordkaper*), the Danes, Norwegians, English, and Americans."[*] The cervical vertebræ in the British Museum, which form the type of Gray's *Halibalæna britannica* are also believed to belong to this species.

Dr. Gray did not recognize *Balæna biscayensis* as a good species, and accounted for the absence of the Right-Whales, formerly found in British waters, from the disturbed state of the seas, owing to the great increase in traffic of ships, and especially steam-vessels, which, he said, "appears to restrict their visits, and especially their breeding, more to the Arctic portion; thus some Whales, which were formerly said to be common on the coast of Britain, as the Right-Whales, no longer visit this country." Eschricht, however, as before stated, has clearly shown that the habits of the northern Right-Whale and localities frequented by them have remained unchanged for many years, as proved by the record kept at the whaling-stations established by the Danish government on the west coast of Greenland.

---

[*] 'Ann. and Mag. Nat. Hist.,' 1878 (11), p. 495.

It is worthy of remark, that in the Southern ocean there are said to be two species of Right-Whale, one *Caperea antipodorum* (Gray), not found further north than 40° south latitude; the other, *Eubalæna australis* (Gray), found as near the equator as 20° south latitude.

The illustration at p. 60 is a reduced copy of the coloured plate in Capellini's account of the Taranto Whale ('*Della Balena di Taranto,*' G. *Capellini, Bologna,* 1877), the original of which was a carefully-executed water-colour drawing, made from the animal itself.

BALÆNOPTERIDÆ.

THE HUMP-BACKED WHALE.

The next family, *Balænopteridæ*, is represented by two genera, *Megaptera* and *Balænoptera.* Like the Right-whales, they all have two blow-holes, but may readily be distinguished by having the throat and belly curiously marked with longitudinal furrows, like the ribs in a worsted stocking: they also possess a well-defined dorsal fin.

The HUMP-BACKED WHALE, *Megaptera longimana* (Rudolphi), the only member of the first genus known to occur in the British seas, has been recorded at least three times; first at Newcastle in September, 1839, again in the estuary of the Dee, in 1863, and in Wick Bay, Caithnesshire, in March, 1871. Capt. Gray tells me they are not uncommon off the east coast of Scotland in summer, and that he has known several captured off Peterhead, three having been brought in in one season. It is possible other examples may have been mistaken for Rorquals, from which this species may at once be distinguished externally by the great length of its flippers, which are white and very conspicuous.

Herr Collett says that this species is met with every spring, on the northern coast of Norway, particularly in the Varanger Fjord; although generally occurring in small numbers, it is occasionally found in great quantities. On one occasion a steam vessel was surrounded by them as far as the eye could see, and great care had to be used to avoid running against them. South of the polar circle, he says it only occurs in small numbers.* In August, 1880, Capt. Gray saw vast numbers of these Whales about one hundred miles N.E. of Iceland; the sea, he states, seemed to be quite full of them as far as he could see from the mast-head. They were accompanied by a small species of "Finner," with a white band across the fin (*B. rostrata*).

The total length of the animal is about 45 to 50 feet, its baleen is black, and the flippers, which are white and notched at the edge, from 10 to 14 feet in length.

THE COMMON RORQUAL.

To the genus *Balænoptera* belong the Rorquals or Fin-whales, the first species of which is the COMMON RORQUAL, *Balænoptera musculus* (Linn.), the *Balænoptera boops* of Bell's first edition, and *Physalus antiquorum* of Gray. This is a much more active animal than the Right-whale; it is difficult of approach, and, upon being harpooned, such is the velocity with which it shoots through the water that the danger is very great; Scoresby mentions one which took out 480 fathoms of line in about one minute. In addition to this, the whalebone is short and of little value, and the yield of oil small; it

---

* 'Bemærkninger til Norges Pattedyrfauna," p. 100. (Særskilt Afryk af 'Nyt Mag. for Naturvsk") 1876.

Fig. 14. COMMON RORQUAL (*Balænoptera musculus*, Linn.)

is therefore avoided by the whalers, as more dangerous than profitable, and if struck at all, it is most likely a case of mistaken identity. From the port of Vadsö, however, the capture of this, and the species immediately preceding and following, is now successfully effected by means of an explosive shell or harpoon, which kills them at once. This fishery was established about the year 1865, by Herr Svend Foyn, from Tonsberg, and is still very successfully prosecuted, as many as 50 Whales being obtained each summer; they are towed into Vadsö, where the blubber is refined and the carcase made into manure.

The habitat of the Common Rorqual is the temperate Northern seas, from the Mediterranean, which it sometimes enters, to the 70° north latitude, and sometimes even farther north still. Nordenskiöld, in the 'Œolus,' last saw Finners on the 18th May, 1861, in lat. 75°45', the temperature of the water being between 2·50° and 3·8° C., and they were not again seen until the return of the expedition in September, in 78° north latitude, the temperature of the water being then about 3·8° C. He remarks, "It is probable that 'Finners' never live in colder water than this, and that the northern limit of their distribution coincides with sea of this temperature. It has to be kept in view, however, that this boundary line lies several degrees further to the north in summer than in winter."*

The range of this group is very great, and, according to Andrew Murray, it would appear that one or more of the Balænopteridæ is found over the whole world, although it is by no means certain that any particular species has a very wide geographical range. *Megaptera longimana*, which occurs in the North Sea, was also supposed to have been met with at the Cape, but Dr. Gray has pointed out differences in the cervical vertebræ of an individual from that locality, which he considers constitute distinct specific characters; on the

---

* 'Arctic Voyages of Adolf Erik Nordenskiöld,' 1858-1879, pp. 51-2.

other hand, a Fin-whale from Java so closely resembles our *Balænoptera laticeps* that Professor Flower, after the most careful examination and comparison almost bone by bone, hesitates to pronounce it distinct, and only separates it provisionally. In our own seas this species is of frequent occurrence, more especially on the Scotch coast, where it appears in the early autumn, attracted by the shoals of herring which abound there at that season. In feeding, the Rorquals are not so restricted to minute marine animals as the Right-Whale, but devour large quantities of fish of various sizes, from herrings up to cod. In the stomach of the Newcastle Humpbacked-Whale (the species mentioned immediately before the present one) were found six cormorants, but a seventh, found in its throat, was supposed to have caused its death by choking it. The blowing is accompanied by a loud noise, which, on a still night, may be heard at a considerable distance. It was formerly supposed that in "blowing" the Whale ejected from its nostrils a very considerable quantity of water, which might be seen to spout up into the air like a fountain; and in the performance of this remarkable feat they were generally depicted. Beale, however, in his 'Natural History of the Sperm Whale,' as early as 1838, showed that this is not the case, and the truth of his observations is now generally acknowledged. The power so to eject water taken into its capacious mouth is, of course, impossible, the blow-hole being in direct communication with the lungs, and not with the cavity of the mouth, nor would it be of any service to the Whalebone-Whales, as the very purpose of the baleen is to form a screening apparatus through which the water is ejected, leaving its minute prey behind; and in the toothed Whales it would not be required. What appears like a jet of water is, in reality, dense vapour—in fact, the breath issuing from the lungs of the animal, highly charged with moisture, which becomes condensed upon exposure to the atmosphere. It often happens, too, that the Whale lets off the imprisoned air just before the blow-hole reaches the surface of the water, or that a wave passes over it at the moment of respiration, the water is thus dashed aside by the blast, and, probably, some of it really carried up into the air, thus heightening the deceptive effect.

This species, when adult, reaches the length of about 70 feet, the upper part is black, the throat and belly white and plaited, the flippers black. The baleen is short and slate colour, veined with streaks of darker shade, but growing lighter towards the inner edge.

Dead Whales, when stranded on the shore, after floating long at sea, are generally greatly distended with gas, which generates rapidly in the tissues after decomposition has set in; in such an inflated condition only a very imperfect conception can be formed of the true proportions of the vast beast. There is frequently, also, a great protrusion of membrane from the mouth, arising from the same cause, and other appearances in the male animal, due to the pressure of gas in the abdominal cavity are generally faithfully portrayed in old books of Natural History.

A Whale of this species, taken off the North coast of Scotland, in April, 1880, was purchased by an enterprising individual in Birmingham, to which town it was conveyed by rail, and there exhibited: probably, this was the greatest distance from the sea at which an entire Cetacean, 63 feet in length, had ever been seen.

The figure of this species is copied, by kind permission of Professor Flower, from the illustration to his paper in the 'Proceedings of the Zoological Society of London' for 1869, p. 604, *et. seq.*

SIBBALD'S RORQUAL.

SIBBALD'S RORQUAL (*Balænoptera sibbaldii*, J. E. Gray; also *Sibbaldius borealis*, Gray, and *Physalis latirostris*, Flower), has several times been met with in British waters, particularly on the east coast of Scotland. It is the largest of this gigantic family, measuring from 80 to perhaps 100 feet in length. One seen by Herr Foyn he estimated at the enormous length of 133 English feet! The famous "Ostend Whale," which was found floating dead in the North Sea, in 1827, and taken into Ostend, belonged to this species; its skeleton was long exhibited in this country, and afterwards in America. Dr. Gray says it is now in St. Petersburg, and gives the total length as 102 feet; as, however, several of the vertebræ are missing, the exact length is uncertain. Professor Turner gives the length of a specimen stranded in the Firth of Forth as 78 feet 9 inches, and the girth behind the flippers about 45 feet: this animal was gravid, but notwithstanding this fact, the bulk must have been enormous.

Herr Rt. Collett, in his 'Norges Pattedyrfauna,' gives a very full account of this species, as observed by him on the Norwegian coast. In June, 1874, he had the opportunity of visiting Herr Svend Foyn's establishment for whale-catching, at Vadsö, and in addition to being enabled to examine three individuals of this species in a fresh state, received much information as to their habits from Herr Foyn and the men engaged in the fishery. This Whale, from its colour, is known by the fishers as the "Blue Whale," and appears to have its home in winter in the open seas, between the North Cape and Spitsbergen. By the end of April or beginning of May it approaches the coast, entering the larger Fjords towards the end of the latter month, to feast upon the enormous quantities of *Thysanopoda inermis*, then found there; it is also seen in summer along the coast from Loffoden to the North Cape, and further to the eastward. When the wind is on the land or in any stormy weather, it seeks the open sea. Varanger Fjord is the favourite hunting-ground for this species, and in the last few years the average number taken there has been thirty; in 1874, as many as 42 were taken: it leaves the Fjord, however, should stormy weather set in. No specimen examined by Herr Collett, or Professor Sars, had taken any other food than *Thysanopoda inermis*, and Herr Foyn and his catchers are all of opinion that they do not eat fish. To obtain the little Crustacean on which they feed and which is found congregated in separate masses, the Whale passes backwards and forwards with its mouth open, till the cavity is well filled, it then closes its capacious jaws upon the contents. Herr Collett found two or three barrels of these small crustaceans in the stomach of a Blue Whale which he examined, and was told that a large one would consume as much as ten barrels.

The female appears, as a rule, to be longer than the male; the young are born about the autumn, one appears to be the usual number, but two yonng ones have more than once been seen with the same old female.

This species may be known by its low dorsal fin, black baleen, and long flippers, which are black above and whitish below: this should be borne in mind, as it is not at all improbable that some, at least, of the enormous cetaceans which are occasionally reported from the North of Scotland, belong to this species; so very unsatisfactory, however, are the reports which appear in print, that it is rarely a single feature is mentioned by which the species may be determined.

RUDOLPHI'S RORQUAL.

RUDOLPHI'S RORQUAL (*Balænoptera laticeps*, J. E. Gray) is a small species which may readily be mistaken for the Lesser Rorqual. A Whale stranded at Charmouth in February, 1840, and described by Mr. Yarrell, under the name of *Balænoptera boöps*, in the proceedings of the Zoological Society for that year, is believed to have been of this species, but the skeleton, although prepared at the time, is supposed to have been sold and converted into manure. The same individual is recorded under the name of *B. tenuirostris*, in the Mag. of Nat. History, iv., 1840, p. 342, by Mr. R. H. Sweeting. Very little is known about the history or distribution of this species; the flippers are entirely black above, wanting the white band found in the next species, and the baleen is believed to be black.

LESSER RORQUAL.

The next and last of the Whalebone-Whales which we know to have occurred in the British Seas is the LESSER RORQUAL (*Balænoptera rostrata*, Fab.; *Rorqualus minor*, Knox), (Fig. 15). Many individuals of this species have been obtained on various parts of the coast, from Cornwall to the North of Scotland. On the coast of Norway it is frequently met with, and is there called the "Bay-Whale," from its habit of entering bays and estuaries; this habit the natives take advantage of for its destruction. Stretching a strong net across the inlet. they cut off its escape, and put a cruel and often protracted end to its existence with harpoons and arrows, the poor Whale sometimes lingering from eight to fourteen days. This species is also known as the "Summer-Whale," and does not appear to be so strictly a northern species as the Balænopteridæ generally are: it is believed, like the Common Rorqual, to have been taken in the Mediterranean. A Whale of this species, taken at Mevagissey, in Cornwall, at the end of April, 1880, was conveyed to London, and there exhibited in the Old Kent Road.

The Lesser Rorqual, from its small size (not exceeding 30 feet), is not liable to be mistaken for any other species except the preceding (Rudolphi's Rorqual), and from that it may be distinguished by the broad white band across its black flipper; the baleen also is nearly white, which is another good

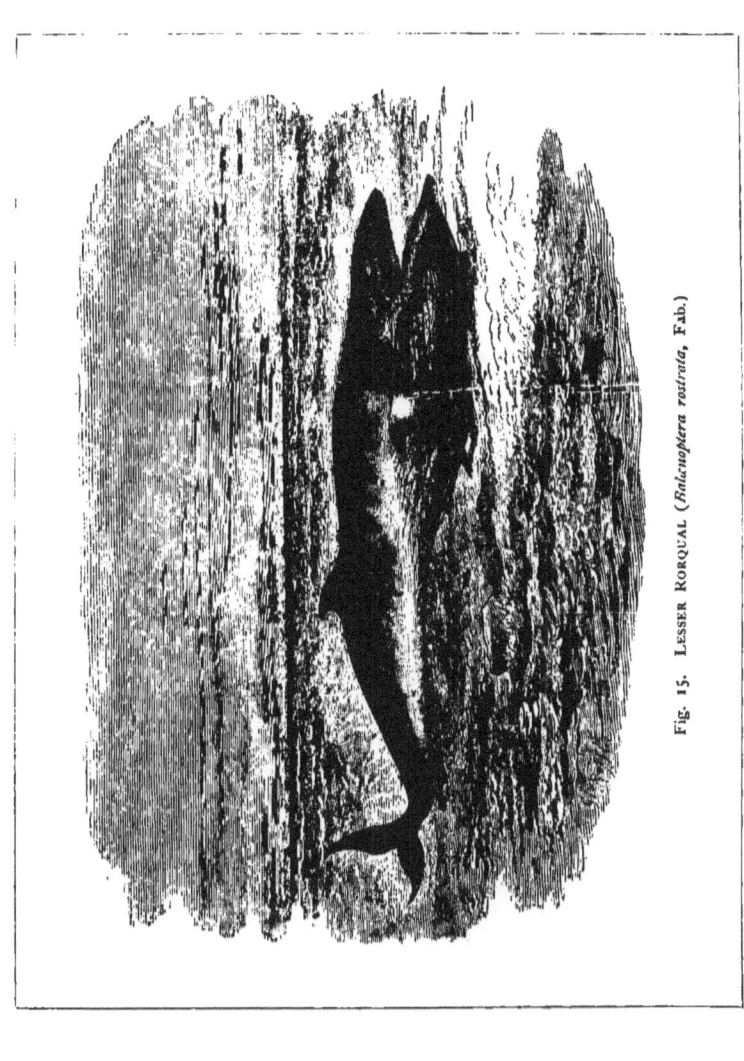

Fig. 15. LESSER RORQUAL (*Balænoptera rostrata*, Fab.)

distinction. The figure of this species is copied from the illustration to an article by Messrs. Carte and Macalister, on the Anatomy of *Balænoptera rostrata*, in the 'Philosophical Transactions' of the Royal Society for 1868, vol. clviii.

In the table on the next page I have endeavoured to give the most striking external peculiarities of our British *Mystacoceti*. They are easily remembered, and will be useful in identifying specimens, should no authority be at hand. The table also indicates the external points to be observed by a person not acquainted with this class of animals, and is most serviceable to enable others to identify doubtful specimens.

TABLE OF DIFFERENCES OF BRITISH MYSTACOCETI (WHALEBONE WHALES).

Species.	Colour. Upper Part.	Colour. Under Part.	Belly and Throat.	Flippers.	Dorsal Fin.	Baleen. Length.	Baleen. Colour.	Total Length.
Balæna mysticetus, Greenland Right-Whale	Dark grey	Throat white	Smooth	Black	None	Long and narrow; 10 or 12 feet	Blackish grey	50 or 60 feet
Balæna biscayensis, Atlantic Right-Whale	Uniform black	Uniform black	Smooth	Black	None	Shorter than the above		40 feet (?)
Megaptera longimana Humpbacked Whale	Black	Black and white	Plaited (plicæ)	Wholly white, about 12 ft. long and notched at the edge	Very low	Short	Black	About 50 feet
Balænoptera musculus, Common Rorqual	Black	White	Plaited	Black	Distinct	Short	Slate colour — shaded lighter to inner edge	About 70 feet
Balænoptera sibbaldii, Sibbald's Rorqual	Black	Slate grey	Plaited	Dark above, White beneath, 12 feet or more long	Very low	Short	Rich black	80 to 100 feet
Balænoptera laticeps, Rudolphi's Rorqual	Black	White	Plaited	Upper part black	?	Short	Black (?)	30 or 40 feet
Balænoptera rostrata, Lesser Rorqual	Black	White	Plaited	Black, with broad band of white across	High	Short	Yellowish white	25 to 30 feet

Fig. 16. Sperm Whale (*Physeter macrocephalus*, Linn.)

ODONTOCETI (TOOTHED WHALES).

PHYSETERIDÆ.

The second sub-order into which the Cetacea are divided, is the *Odontoceti*, or Toothed Whales. In this section, baleen is never present, but well-developed teeth are found in one or both jaws of the adult; in some species they are very numerous; sometimes, though rarely, deciduous. The blow-hole is single, and the skull generally asymmetrical, or not precisely alike on both sides of the medial line. Professor Flower divides the *Odontoceti* into three families, one of which, the *Platanistidæ*, as already said, is found only in India and South America: the other two, *Physeteridæ* and *Delphinidæ*, are represented in our Fauna by about fifteen species.

Of the *Physeteridæ*, four genera are represented in the British fauna by four or five species; namely, one *Physeter*, the Sperm Whale; two *Hyperoodons*, the common Beaked Whale, and a doubtful species called the Broad-fronted Beaked Whale; one *Ziphius*, Cuvier's Whale; and one *Mesoplodon*, Sowerby's Whale.

SPERM WHALE, OR CACHELOT.

By far the most conspicuous species of this interesting group is the SPERM WHALE, *Physeter macrocephalus* (Linnæus), which rivals the Right-Whale in commercial importance, and in the value of its products. This species has a very wide geographical range, having been found in almost every sea between lat. 60° north and 60° south. The attempt has been made, I think unsuccessfully, to show that the Sperm Whale of the Southern Hemisphere is distinct from that of the northern; there seems, however, no reason, at present, to doubt, although, of course, it may eventually be found otherwise, that the same species of Sperm Whale ranges over the whole of this vast tract of ocean. North of about 40° it appears to be only a straggler, and although the Arctic seas are almost always stated by authors to be its head-quarters, very few well-authenticated instances of its occurrence farther north than Scotland are on record; Lilljeborg excludes it from his account of the Scandinavian cetacea, but Herr Collett says that within the last 100 years, at least two individuals of this species have been stranded on the Norwegian coast, and that Professor Sars, during a stay in Loffoden, received information which convinced him that one was seen there in the summer of 1865.

From the middle of the sixteenth to the middle of the seventeenth century, the stranding of individuals of this species on the coast of Great Britain, and, indeed, of other countries in Europe from the Netherlands to the Mediterranean, was by no means a rare occurrence; these were generally solitary males, but occasionally small "schools" were met with, as in July, 1577, in the Scheldt, where three were taken; also, at Hunstanton, in Norfolk, in 1646, mentioned below.

Of its occurrence on the British coast there are numerous instances; in all cases, however, they are believed by Andrew Murray to have been stragglers, "which have rounded Cape Horn (they have never been known to double the Cape of Good Hope) or unpromising colonies, for they are becoming scarcer and scarcer in more than their due proportion."* Eight or ten individuals of this species have occurred on the coast of Scotland between the years 1689 and 1871 (Alston, 'Fauna of Scot.', p. 18).

---

* 'Geographical Distribution of Mammalia,' by Andrew Murray, 1866, p. 211.

In the church of St. Nicholas, at Great Yarmouth, is the basal portion of a skull of this animal, which has been converted into a chair : it formerly stood outside the church, and of course, as it was an object of wonder, it was relegated to the powers of darkness, and *christened* (?) the "Devil's Seat ; " it has, however, now been admitted into mother church, and stands beside the north-west door under the clock. In the churchwardens' accounts for 1606 there is a charge of 8s. for painting this chair, which clearly proves its antiquity. In a letter to Sir Thomas Browne (Wilkins' edit., 1852, editor's preface to "Pseudodoxia," vol. i. p. lxxxi.), Sir Hamon L'Estrange writes

Fig. 17. Chair in Great Yarmouth Church, formed from the basal portion of the skull of the Sperm Whale.

Fig. 18. Back view of the same.

that in June, 1626, a Whale, afterwards referred to by Sir T. Browne as a Sperm Whale (vol. iii. p. 324), was cast upon his shore or sea-liberty, "sometyme parcel of the possessions of the Abbey of Ramsey, &c." The same author, in his account of the "Fishes found in Norfolk and on the Coast," says, " A Spermaceti Whale of 62 feet long [came on shore] near Wells, another of the same kind twenty years before at Hunstanton [the one referred to by Sir H. L'Estrange] ; and not far off, eight or nine came ashore, and two had young ones after they were forsaken by the water." The Whale mentioned by Sir H. L'Estrange came on shore in 1626 ; twenty years after would give 1646 as the date of the Wells specimen; and in December of that year, according to Booth's "History of Norfolk," published in 1781 (vol. ix. p. 33), "A great Whale was cast on the shore here [at Holme-next-the-Sea], the wind blowing strongly at the north-west, 57 feet long, the breadth of the nose-end eight feet, from nose-end to the eye. 15½ feet; the eyes about the same bigness as those of an ox, the lower chap closed and shut about four feet short of the upper; this lower chap narrow towards the end, and therein were 46 teeth like the tusks of an elephant; the upper one had no teeth, but sockets of bones to receive the teeth: two small fins only, one on each side,

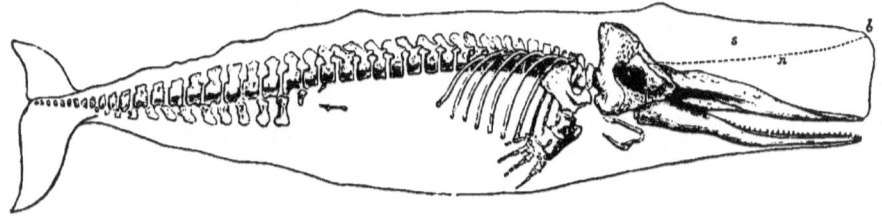

Fig. 19. SKELETON OF THE SPERM WHALE (after Flower).

s, Spermaceti Cavity; n, Nasal Passage, in dotted line; b, Blow-hole.

and a short small fin on the back; it was a male . . . . ; the breadth of the tail, from one outward tip to the other, was 13½ feet. The profit made of it was £217 6s. 7d., and the charge in cutting it up and managing it came to £100 or more." It seems probable that a "school" got bewildered in the shallow waters of the Wash, and that the individual of which Booth gives such an excellent description, formed one of the same party as the eight or nine mentioned by Sir T. Browne. In May, 1652, Mr. Arthur Bacon writes to

Sir T. Browne about the Sperm Whale cast on shore at Yarmouth, but the actual date of the occurrence is not given. Since these ancient records, many others have occurred at intervals, singly or in small parties, on various parts of the coast; the last instance, I believe, being in July, 1871, when one was stranded on the shore of the Isle of Skye.

Of the osteology of the Sperm Whale, Professor Flower has given an exhaustive description in a paper published in the 'Transactions' of the Zoological Society, vol. vi., and of its habits a very interesting account is given by Thomas Beale, who, in the capacity of surgeon on board ships employed in the South Sea fishery, had unusual opportunities of observing this remarkable animal. He published a book entitled 'The Natural History of the Sperm Whale,' to which I am largely indebted for what I shall have to say about this species.

The colour of the Sperm Whale is black above and grey beneath, the colours gradually shading into each other. The full-grown male is about sixty feet long; the females are much smaller and more slender than the males. The head, which constitutes more than one-third of the whole of the animal, presents a very remarkable appearance, the truncated form of the snout looking as though it were cut off at right-angles to the body: at the upper angle is situated the single blow-hole. The juncture of the head with the body is the thickest portion, and the body decreases little in size till the "hump," which is situated in the place of the dorsal fin, is reached; from this point it rapidly diminishes to the tail. The flukes of the tail are from twelve to fourteen feet in breadth, and the two flippers each about six feet long. The under jaw is pointed, and about two feet shorter than the upper; it is furnished with about twenty-five large conical teeth on each side; but the number is not constant, nor is it always the same on each side. In the upper jaw are no visible teeth, but those of the lower jaw shut into corresponding depressions in the upper. The tongue is small, and, like the lining of the mouth, of a white colour. The upper part of the head, called the "case," contains the "spermaceti," which upon the death of the animal granulates into a yellowish substance. Beale says that a large Whale not unfrequently contains a ton of spermaceti. Beneath the "case" is situated the "junk," which consists of a dense cellular mass, containing oil and spermaceti. The blubber is about fourteen inches thick on the breast, and in most other parts of the body from eight to eleven inches. By the whalers this covering is called the "blanket." With regard to the apparently ungainly head of the Sperm Whale, Beale remarks as follows:—"One of the peculiarities of the

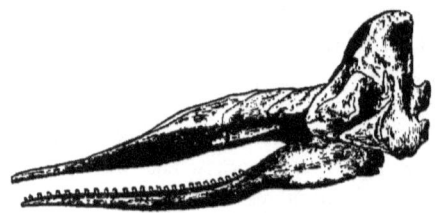

Fig. 20. SKULL OF SPERM WHALE.

Sperm Whale, which strikes at first sight every beholder, is the apparently disproportionate and unwieldy bulk of the head; but this peculiarity, instead of being, as might be supposed, an impediment to the freedom of the animal's motion in its native element, is, in fact, on the contrary, in some respects, very conducive to its lightness and agility, if such a term can with propriety be applied to such an enormous creature; for a great part of the bulk of the head ·is made up of a thin membranous case, containing, during life, a thin oil, of much less specific gravity than water, below which is again the junk, which, although heavier than the spermaceti, is still lighter than the element in which the Whale moves; consequently, the head, taken as a whole, is lighter specifically than any other part of the body, and will always have a tendency to rise at least so far above the surface as to elevate the nostril or "blow-hole" sufficiently for all purposes of respiration; and more than this, a very slight effort on the part of the fish would only be necessary to raise the whole of the anterior flat surface of the nose out of the water. In case the animal should wish to increase his speed to the utmost, the narrow inferior surface, which has been before stated to bear some resemblance to the cutwater of a ship, and which would, in fact, answer the same purpose to the Whale, would be the only part exposed to the pressure of the water in front, enabling him thus to pass with the greatest celerity and ease through the boundless track of his wide domain."* When swimming at ease, the Sperm Whale keeps just below the surface of the water, and goes at about three or four miles an hour; but on an emergency it is able to attain a speed of ten or twelve miles an hour: it then progresses by means of powerful lateral strokes of its tail, and alternately rises and sinks at each stroke. In progressing in this manner, the blunt anterior surface of the head never presents itself directly to the water; the animal's body being in an oblique position, it is only the angle formed by the inferior surface which first presents itself, and this, which Beale likens to the "cutwater" of a ship, offers the least possible amount of resistance.

When undisturbed, the Sperm Whale rises to the surface to breathe about once every hour. Beale says the regularity with which every action connected with its breathing is performed is remarkable; the time occupied differs slightly in each individual, but each one is minutely regular in the performance of every action connected with respiration, so that the whalers know how long it will remain beneath the surface before reappearing to renew its supply of air. A full-grown "bull," he says, remains

---

* 'Natural History of the Sperm Whale,' p. 28.

at the surface ten or eleven minutes, during which he makes sixty or seventy expirations; after which he disappears, to return again to the surface in one hour and ten minutes. The blowing is not accompanied by any sound, and notwithstanding the wonderful accounts of its roarings and bellowings, the Sperm Whale may be said to be an absolutely silent animal. The females and young males are gregarious, but are found in separate herds or "schools," as they are called. A "school" will sometimes consist of five or six hundred individuals. The herds of females are always accompanied by from one to three large "bulls;" but the full-grown males are said to be generally solitary in their habits, except on certain occasions, when it is supposed they are migrating from one feeding-place to another. The majority of those which occur on our coast are these solitary males; when they visit us in herds, as mentioned by Sir Thomas Browne, they are all probably females or young males. The "bulls" are very fierce and jealous, and fight fiercely. The females show great attachment to each other and to their young, so much so that, one being wounded, the others of the herd remain and fall a comparatively easy prey. The young males, on the other hand, are very wary and difficult of approach, and should one be attacked, the others immediately take the alarm and retreat. The female produces one young one, rarely two, at a time, and breeds at all seasons of the year. Their senses of sight and hearing are very acute, and after being once unsuccessfully attacked, they are very difficult and dangerous to approach.

The food of the Sperm Whale consists almost entirely of Cephalopode Mollusks (cuttle-fish), although at times, when feeding near the shore, it has been known to take fish as large as salmon. How it contrives to capture such active prey as fish seems difficult to conceive. Beale is of opinion that the Whale sinks to a proper depth in the sea, where remaining as quiet as possible, and opening wide its mouth, the prey are attracted by the glistening white colour of its lining membrane, curiosity leading them to destruction;

for no sooner have a sufficient number entered his mouth than the Whale rapidly closes his under jaw, and they are made prisoners, and swallowed.

The pursuit of the Sperm Whale is attended with much greater danger than that of the Greenland Whale, and Beale gives many instances in which, in his own experience, boats were stove in and men lost; stories of fighting Whales, he says, are numerous, and probably much exaggerated; one, known as "Timor Jack," is said to have destroyed every boat sent against him, till at last he was killed by approaching him from several directions at the same time, his attention thus being diverted from the boat which made the successful attack. Another fish, known as "New Zealand Tom," destroyed nine boats successively before breakfast, and when eventually captured, after destroying many other boats, many harpoons from the various ships which had attacked him were found sticking in his body. There is one well-authenticated instance of a vessel being attacked and destroyed by a Sperm Whale: the American whale-ship *Essex* was attacked by one, which, first passing under the vessel, probably by accident, came in contact with her keel and carried it away: then turning and rushing furiously upon the ship, the Whale stove in her bow; so serious was the breach that the vessel speedily filled and went down. Most of the crew were away in their boats at the time, but those on board had just time to launch their one remaining boat before the vessel sank. The boats made for the coast of Peru, the nearest land, many hundreds of miles distant; one of them was picked up drifting at sea, and three of the crew, who were found in it in a state of insensibility, were the only survivors of the ill-fated vessel.

In addition to the sperm and oil, this species yields another product which is, or was, very valuable, although it is the result of disease, and one would imagine a very uninviting substance—I refer to *Ambergris*, the origin and composition of which was so long a puzzle to the learned. This substance is now well known to be a concretion of the indigestible portions of the Cuttle-fish, which form the food of the Sperm Whale. The nucleus of the mass is generally the horny beaks of these creatures, and the substance itself is found in the intestines of the Sperm Whale, or on the shores of the seas frequented by this species: no other Whale is known to be subject to these bezoars. It was formerly believed that the origin of ambergris was in some way connected with the sea, and when it was afterwards found in Whales, the fact was simply attributed to their having swallowed it. Sir Thomas Browne writes of the Sperm Whale which came on shore at Wells, in 1646:—"In vain was it to rake for ambergriese in the paunch of this leviathan, as Greenland discoverers and attests of experience dictate that they sometimes swallow great lumps thereof in the sea; insufferable fœtor denying that inquiry; and yet if, as Paracelsus encourageth, ordure makes the best musk, and from the most fœtid substances may be drawn the most odoriferous essences; all that had not Vespasian's nose (*Cui odor lucri ex re qualibet*) might boldly swear here was a subject fit for such extractions" (vol. i., p. 356). It was not until 1783, in a paper read before the Royal Society by Dr. Swediaur, that a scientific account of the origin of ambergris was made known. At the present time its medical virtues, which were formerly considered very great, are altogether at a discount, and the only use to which it is applied is in the preparation of perfumery.

The South Sea whale-fishery was long prosecuted by the Americans before the British ships took part in it, from 1771 to 1775 Massachusetts is said by McCulloch to have had 121 vessels in this trade; about the beginning of the American war, however, the English also sent out ships, and in 1791 had 75 vessels engaged in the South Seas. The number of British ships, as with those employed in the northern fisheries, varied considerably, influenced probably by the varying amounts of bounty offered by the Government, but never exceeded 75; in 1815 they had fallen off to 22; in 1820 they again rose to 68, from which they gradually fell to 31 in 1829, all of which sailed from the port of London. Beale sailed from London, in 1831, in the "Kent," returning in the "Sarah and Elizabeth," both of which vessels belonged to Thomas Sturge. The duration of the voyage was from two to four or even five years, the average of 199 voyages being three years and three months, and the yield of oil, 169 tons per voyage. At the present time no British vessels are engaged in the South Sea trade, which has again reverted to the Americans.

I have said very little about the method of pursuit and capture of this species, and of the Right-Whale, because it is a subject in which I take no pleasure; those who wish to know how these peaceful and highly-organised giants are approached, and how they behave when terrified and smarting under the harpoon and whale-lance, can pursue the subject *ad nauseam* in the pages of Scoresby, Beale, and others; the sickening process of "flensing" and disposing of the blubber is described with equal minuteness. The halo of romance with which some authors seek to surround the whale-fishery, is, doubtless, in a great measure due to the solitary and distant fields of operation, whether it be in the frozen regions of the north, or the vast and trackless oceans of the south, but its stern reality is prosaic enough. The occupation is one of hardship and danger, but the remuneration when successful is large in proportion, and I can hardly conceive, under any circumstances, of men inflicting the fearful amount of suffering which every "full" whale-ship, or in a still greater degree every "full" sealer, represents. Science is constantly adding to our resources, and it is sincerely to be hoped that ere long substitutes may be found for animal oil and whalebone which will supersede their use in the few processes in which they are still requisite: should this be long delayed, it is to be feared that the Seals and Whales, at least of the northern seas, will soon cease to exist. In the meantime, it is gratifying to find that it is from the sealers and whalers themselves that the demand for the better regulation of the trade has emanated, and the name of

Captain David Gray, of Peterhead, stands prominent amongst those who have urged upon the governments of this and other countries concerned, such regulations as shall insure greater humanity in its prosecution, and prevent the wasteful destruction which, if continued, must speedily ruin a valuable source of commercial enterprise.

Although so widely spread over the waters of the globe, possessing, I believe, a range greater than any other known mammal, it is only open and deep waters which can be said to be the home of the Sperm Whale; when found in shallow seas, its generally emaciated condition indicates the absence of its proper nourishment; and the readiness with which whole herds precipitate themselves stupidly upon the sands, shows how little they are acquainted with such objects. Mr. Andrew Murray makes some observations upon this subject, which are so interesting and so suggestive that I cannot resist making a long quotation.

Speaking of those specimens which have now and then been cast ashore in the North Atlantic or in the English seas, he says: " They seem to be unprepared for, or not adapted for, shallow seas. Accustomed (perhaps not individually, but by hereditary practice or instinct) to swim along the coral islands of the Pacific within a stone's throw from the shore, they cannot understand, their instinct is not prepared to meet, shallow coasts and projecting headlands. If they were habitual residents in our seas, they must either be speedily extirpated, learn more caution, or be developed into a new species." .... Mr. Murray further says: "I observe that almost every place that has been above mentioned as a favourite resort of the Sperm Whales, although not out of soundings, has claims to be considered the site of submerged land. The islands in the Polynesia, which are its special feeding-ground, are the beacons left by the submerged Pacific continent. In pure deep seas animal life is usually scarce, and the absence of breeding-ground is probably the chief cause of it; but this only applies to a certain kind of animals, those which require a bottom on which to deposit their spawn; but there are many which do not require this. The spawn of some floats about unattached; for others a frond of weed is sufficient attachment; and it has occurred to me that the distribution of the Sperm Whale may in some way be connected with the geological antecedents of the ocean it inhabits. I think it not improbable that the site of a submerged land may swarm with life, which originally proceeded, or was dependent on it, long after it had been in the deep bosom of the ocean buried. The Sargasso seas, which swarm with *Eolidæ* and *Crustacea*, are examples of this life; it is not invariably either present or absent in deep water, and it is its presence or its absence which is instructive. Those animals which required a bottom to spawn upon may have died out or been developed into others which do not; and those which do not require such a support may have multiplied correspondingly. In one of the maps in Lieutenant Maury's book, already cited, there is a space of sea opposite the western coast of South America, and lying between Patagonia and New Zealand, marked 'Desolate region, distinguished by the absence of animal or vegetable life';—no Sperm Whales here—nothing for them to feed upon—and no symptoms, either by banks of Sargasso or coral islets, of any land ever having existed there. There is no apparent reason why this place, except from some special cause peculiar to itself, should be more desolate than any other in the same latitude—than the deep sea on the east side of Patagonia, for example. I can imagine that, if the bottom of the sea should subside gradually, where animal life had once abounded, animal life—not that animal life, but animal life due in some way to it—might continue to linger over it long after it had passed beyond the depth at which it could practically have any effect upon the animal life above it; but if a part of the circumference of the globe has always been under water, before and ever since the creation of life, no life is likely to be found on that spot, because it has never had a starting-point of life from which to begin; and, as already said, a slender barrier stops the spread of species, and species would certainly not spread to a spot where there was nothing for them to feed upon. Again, animal life could not begin to feed upon animal life till vegetable life had previously prepared the way, by providing food for the animals which were to furnish food for others.; and vegetable life could not begin to grow without a foundation of land, accessible either above or below water. The total and constant absence of all life at any particular spot appears to me, therefore, to furnish a presumption that there has never been dry land or shallow water there. Whether the continuance of deep water in one spot for some interminably long time might not have the same effect is another question, which, whatever way it may be answered, would not affect my explanation of the cause of the absence of the Sperm Whale from such spots."[*]

The woodcuts (figs. 17 and 18), representing the chair in Yarmouth Church, which is formed of part of the skull of an individual of this species, are from the 'Purlestrations of Great Yarmouth,' by Mr. C. J. Palmer.

THE ZIPHIOID WHALES.

The sub-family *Ziphiinæ*, which follows next, is, perhaps, the most remarkable of the whole of this interesting order. The *Ziphioid* Whales, as they are designated, are, for the most part, very rare, and until the commencement of the present century, with one exception, were known to science only from their numerous remains, found chiefly in the Crag deposits. Even so recently as 1871, Professor Flower, in a memoir of this group[†] speaks of their occurrence at irregular intervals, and at various and most

---

[*] 'Geographical Distribution of Mammalia,' pp. 211-13.
[†] 'Transactions' of the Zoological Society, viii., p. 203.

distant parts of the world, to the number of about 30 individuals, in all cases solitary, and that their habits were almost absolutely unknown. Since that time, however, very considerable additions have been made to our knowledge of the group, and Professor Flower, in a second contribution on the same subject* made in 1877, states that "instead of being so rare as was then supposed, since the attention of naturalists resident in our colonies has been directed to the importance of losing no opportunity of securing such specimens as accidents of wind and waves may cast upon their shores, it has been proved that in the seas of the Southern Hemisphere these Whales exist in considerable numbers, both as species and as individuals, and that one species, at least [*Mesoplodon grayi*] is gregarious, having been met with in two instances in 'schools' of considerable numbers." "The geographical distribution of the group," adds Professor Flower† "has a very great interest in relation to that of many other Australian groups, both of vertebrates and invertebrates. Among the earliest known remains of Cetacea, in the Belgian and Suffolk Crags, *Mesoplodon* and closely-allied forms are most abundant. Up to a little more than ten years ago, the few stray individuals of *Mesoplodon bidens* occasionally stranded on the shores of North Europe, were supposed to be their sole survivors. Since that time it has been proved that they are still numerous in species, and even in individuals . . . . . . in the seas which surround the Australian continent, extending from the Cape of Good Hope on the one side, to New Zealand on the other, though beyond these limits no specimens have yet been met with. It is the history of the Marsupial Mammals, of *Ceratodus*, of *Terebratula*, and of numerous other forms."

The group is divided into four genera—*Hyperoodon, Berardius, Ziphius,* and *Mesoplodon* (the second of which is not represented in our Fauna). Its

---

* *Ibid.* x., p. 415.   † *Ibid.*, p. 435.

members were formerly distinguished by the absence of functional teeth in the upper jaw, but, recently, a row of small teeth, of determinate number and definite form, has been discovered in many individuals of a species of *Mesoplodon*. The teeth in the lower jaw are always quite rudimentary, with the exception of one, or occasionally, two pairs. These may be largely developed, especially in the male sex, and are placed, generally, well forward. "They have a small and pointed enamel-covered crown, composed of true dentine, which, instead of surmounting a root of the ordinary character, is raised upon a solid mass of osteo-dentine, the continuous growth of which greatly alters the form and general appearance of the organ as age advances." In *Mesoplodon layardi* this little dentine cap is not larger than the portion of the tooth ordinarily shown above the gum, but the fang-like growth is so great that the tips of the "tusks" meet over the upper jaw, so that the animal is only able to open its mouth for a very short distance indeed. The form assumed in *Mesoplodon bidens* will be seen in the figure of the head of that species, at p. 104. The blow-hole is sub-crescentic, and a pair of remarkable furrows occurs in the skin of the throat, almost in the form of the letter V, the point directed forward. The skull presents a remarkable appearance in the genus *Hyperoodon*, caused by the enormous maxillary crests which produce the peculiar conformation of the head in the living animal, originating the trivial name "Bottle-head." The food of the whole group is said to consist mainly of *Loligo*, commonly called "Squid," and other Cephalopods which frequent the open sea.

One very singular circumstance with regard to these creatures is that they never seem to be taken at sea, but, whenever procured, it is by their running themselves on shore. This, as before remarked with regard to the Sperm Whale, would seem to indicate that their natural habitat is the deep waters of the open seas, where shallows are unknown. The sand-banks which surround a sloping shore, of which they have had no experience, speedily prove fatal to them.

BEAKED WHALE.

The common BEAKED WHALE, or BOTTLE-HEAD (*Hyperoodon rostratum*, Chemnitz ; *Hyperoodon butzkopf*, Lacepède), is of frequent occurrence in the North Atlantic, and generally visits our shores in autumn, sometimes ascending the estuaries of rivers: it has been taken several times at the entrance to the river Ouse. It is solitary in its habits, more than two being never met with in the same place, and in that case it is often the old female and her young one: the old male is said to be very shy and rarely secured. In September, 1877, an adult female, 24 ft. long, was taken in the Menai Straits; it was accompanied by another, probably its young one. Capt. Feilden met with what he believes to have been this species, just within the Arctic Circle; "each emission of breath was accompanied by a stentorian grunt, which closely resembled that of an elephant."[*]

The colour is black above, the under parts being lighter: the two teeth in the lower jaw are generally hidden in the gum. Its food consists of cuttle-fish, the remains of great numbers of which have been found in its stomach.

BROAD-FRONTED BEAKED WHALE.

Another species of *Hyperoodon*, for which the name *H. latifrons* has been proposed, is by some supposed to exist. Scarcely anything is known about it as a species. "The principal distinctive characters of the skull lie in the great raised crests of the maxillary bones, which are very much thickened

---
[*] *Zoologist*, 1878, p. 319.

and flattened above, so as almost to touch one another, whereas, in *H. rostratum*, they are rather sharp-edged above, and separated by a considerable interval. In *H. latifrons*, these crests rise absolutely *higher* than the occipital region of the skull, which is not the case in the common species."* Individuals possessing these peculiarities have been taken three or four times on the British coast, and on one occasion, in Greenland. Another was stranded in 1873, at Hasvig, near Hammerfest, and identified by Professor Sars from its remains; its length was 30 feet (Norse), and the colour dark on the back, but lighter beneath.† It has, however, been suggested, with much probability, by Eschricht, that these individuals are, after all, only the males of the preceding species; for all the specimens with broad crests, of which the sex was noted, were males.

CUVIER'S WHALE.

CUVIER'S WHALE (*Ziphius cavirostris*, Cuv.; *Epiodon desmarestii*, J. E. Gray, 'Cat. Seals and Whales'), another of this remarkable group, has been met with once on the coast of Shetland, and it, or its remains, have been found about five or six times in other parts of Europe, and also, it is believed, at the Cape of Good Hope, the east coast of South America, and New Zealand. Professor Turner is of opinion that the geographical range of *Ziphius cavirostris* equals that possessed by the Spermaceti Whale.‡ In colour this species is believed to resemble Sowerby's Whale; it has two teeth, one on each side of the lower jaw, close to the extremity.

---

* Bell's 'Brit. Quad.' p. 426. † Collett, 'Norges Pattedyrfauna,' p. 99.
‡ 'Zoology of H. M. S. Challenger,' part iv., p. 29.

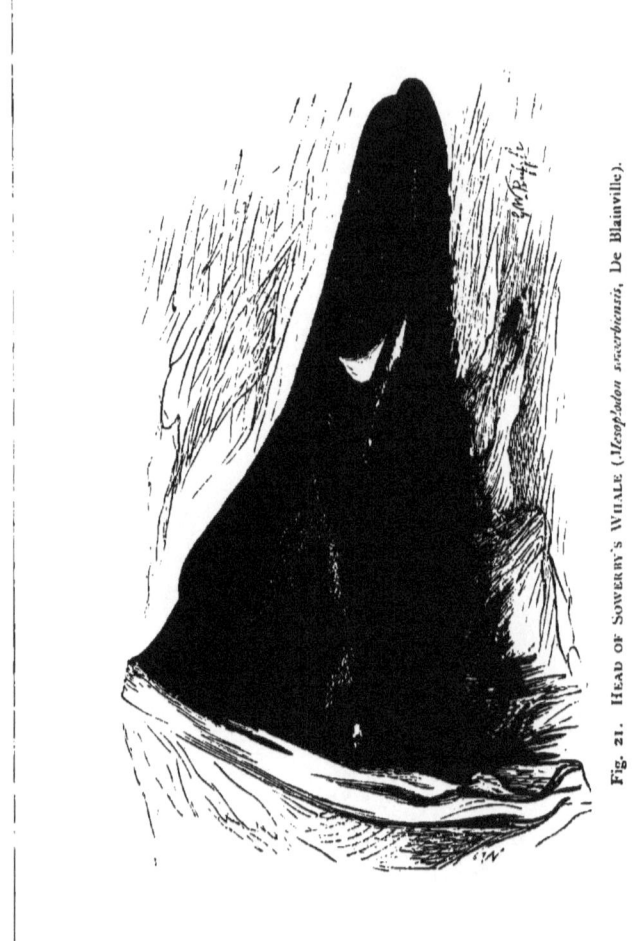

Fig. 21. Head of Sowerby's Whale (*Mesoplodon sowerbiensis*, De Blainville). From Trans. Roy. Irish Acad.

Cuvier established the genus *Ziphius* in 1825, from a fossil skull found on the coast of Provence in 1804, which he believed at the time to belong to an extinct animal.

SOWERBY'S WHALE.

One more British Ziphioid is known, SOWERBY'S WHALE (*Mesoplodon sowerbiensis*, De Blainville); it was first described from a specimen which came ashore at Brodie, Elginshire, in 1800, and has since been found three times in Ireland; there is also a skull in the Museum of Science and Art at Edinburgh, which belonged to a specimen believed to have been captured somewhere on the Scotch coast; the remains of five others are preserved in various Continental museums.

Of the individual which came on shore on the coast of Kerry, in March, 1864, Mr. Andrews has given a description in the "Transactions of the Royal Irish Academy," for April, 1867. Fortunately, it came under the notice of Dr. Busteed, of Castle Gregory, who being interested in zoology, and aware of the great importance of the occurrence, photographed the head in several positions while it was yet fresh: Dr. Busteed's photographs were reproduced in the Transactions of the Royal Irish Academy. The head had unfortunately been removed immediately behind the frontal portion of the skull, the base of which is lost, as are also the other parts of the skeleton. The total length of the animal was about fifteen feet, the two teeth largely developed and projecting like the tusks of a boar. On the under part of the throat the V-shaped furrow was very conspicuous. Sowerby's specimen was coloured black above, and nearly white below. The skin was smooth like satin. "Immediately under the cuticle the sides were completely covered with white vermicular streaks in every direction, which at a little distance appeared like irregular cuts with a sharp instrument."

DELPHINIDÆ.

The remaining family, *Delphinidæ*, as before stated, is a very numerous one. It has ten representatives in the British fauna, contained in seven genera, the first of which, according to the arrangement I have adopted, is that of *Monodon*.

THE NARWHAL.

The NARWHAL (*Monodon monoceros*, Linn.) is a native of the Polar seas seldom leaving the ice; stragglers have occurred three times on the British coast, one in 1648 in the Firth of Forth, another came ashore alive at Boston, in 1800; the third was taken in Shetland in 1808.

This species is very numerous in the frozen seas to the north of latitude 65°, and is remarkable for the enormous development in the male of the left canine tooth, which is projected forward in the form of a tusk or spear, reaching to the length of six or eight feet, while the right tusk remains abortive, and does not pierce the alveolus. The spear is of fine compact ivory, hollow for the greater part of its length, grooved spirally from left to right, along its outer surface, the spiral generally making five or six turns, but smooth at the end, and bluntly pointed. Although the right canine is rarely developed, a few examples have occurred in which both tusks were present; the female is very rarely furnished with this appendage.

Mr. J. W. Clark, in a paper on a 'Skeleton of Narwhal, with two fully-developed tusks,'[*] writes as follows:—"The skulls of the Toothed Whales

---

[*] *Proc. Zool. Soc.*, 1871, pp. 41-53.

are generally asymmetrical, being twisted more or less, usually towards the left. This peculiarity is especially observable in Monodon. One would expect it to be greatly exaggerated in the skulls of the males, where the left tusk alone is developed, and the left maxillary is, in consequence, very large, and the right proportionately small; but it does not seem to be affected by the absence or presence of the teeth. Female skulls, where neither tusk is developed, are equally twisted, and so are the bidental skulls . . . . the increased size of the right maxillary does not appear to affect the rest of the skull."

Mr. Clark enumerates eleven skulls of the Narwhal in which both tusks are developed; four at Copenhagen, and one each in the museum of Hamburg, Christiania, Amsterdam, Weimar, Hull, Paris, and Cambridge; to these must be added a twelfth, which was brought from Prince Regent's Inlet, by Capt. Gravill, of the "Camperdown," and is now in the Dundee Museum.

Not long since I saw preserved in a country mansion, the tusk of a Narwhal measuring 7 ft. 5 in. long; it was carefully kept in a long case resembling a barber's pole, and bore a ticket attached, which stated that it was "Bequeathed in 1561 by the Countess of ———, to her daughter ———." No doubt at the time this formed a valuable bequest, as even royal and ecclesiastical dignitaries are said to have esteemed these strange objects (probably associated with the mythical unicorn), as "good against" poisons and fevers, and prized them accordingly. The use of this remarkable appendage appears very doubtful; it has been conjectured that it serves to stir up food from the bottom of the sea, in which case the female would be badly off without it; or that it is employed to keep breathing-holes open in the ice, and an instance is related in support of this view, in which hundreds were seen at an ice-hole protruding their heads to breathe, but it is not clear whether they made the hole for themselves, or whether they were attracted by it, particularly as there were numbers of White Whales with them. It seems certain, however, that the tusk, which is frequently found in a broken condition, is used for purposes of attack and defence. Like the horn of the stag, it is, no doubt, a sexual distinction.

The Narwhal is very social in its habits, great numbers being often met with together; its food consists of cuttle-fish and crustaceans. The length of the full-grown animal is about 16 feet, the upper parts gray, the sides and belly white, and the whole animal spotted with black and gray. The only authentic figure of the Narwhal with which I am acquainted is that given by Scoresby; this is so well known from frequent reproduction that it is not necessary to give it here.

THE WHITE WHALE.

The WHITE WHALE, or BELUGA (*Delphinapterus leucas*, Pallas), like the preceding species, is a native of the Polar seas, where it is common; it is abundant in the White and Kara Seas, and in the Gulf of Obi; on the coast of Norway it is occasionally met with. From Scotland, five individuals have been recorded, but it must be regarded as only an accidental straggler. On the east coast of America it is found as far south as the Gulf of St. Lawrence, where, as in the White Sea, it delights in ascending the mouths of large rivers.

No English examples have been met with, but, in the British Association Report on the Fauna of Devonshire (1869, pp. 84 and 85) occurs the following passage. "Mr. H. P. Gosse writes:—'On August 5th, 1832, I was returning from Newfoundland to England, and was sailing up the British Channel close to the land, when, just off Berry Head, I saw under the ship's bows a large cetacean of a milky white hue, but appearing slightly tinged with green from the intervening stratum of clear water. It was about 16 feet long, with a round bluff head. It continued to swim along before the vessel's head, a few yards beneath the surface, for about ten minutes, maintaining our rate of speed, which was five knots an hour, all which time I enjoyed from the bowsprit a very good view of it. It could have been no other than the White Whale, the *B. borealis* of Lesson.'" Mr. Alston also states that Mr. J. G. Gordon informed him that in June, 1878, "he saw a large white cetacean, presumably of this species, in Loch Etive."

Fig. 22. BELUGA, caught by the tail, near Dunrobin, Sutherlandshire.

In a communication to the Zoological Society of London,* quoting a letter from the Rev. Dr. Joass, of Golspie, Professor Flower thus describes the singular capture of one of these rare visitants to our seas:—" It was found close to the salmon-nets, near the Little Ferry, about three miles to the westward of Dunrobin, Sutherlandshire, at ebb tide, on Monday, June 9th, 1879, caught by the tail between two short posts, to which a stake-net was fastened; and a salmon, of 18 lbs. weight, which was supposed to have been

---

* *Proc. Zool. Soc.*, 1879, pp. 667-9 (by which Society the above woodcut was kindly lent).

the object of its pursuit, was found in front of it. It measured 12 ft. 6 in. in length. The tail was 34 inches across, and the flippers 17 inches long. It was a female [adult] and had twenty teeth in the upper jaw, and sixteen in the lower. The stomach contained a few flakes of fish, which, from their size and colour, might have been salmon. . . . . . I have heard since, that two days before its capture, it was seen off Cracaig by Brora fishermen, who were lying at their lines. At first they thought it was a human body; as it approached, *against the ebb*, they took it for a ghost!" On examining the skull of this specimen, Professor Flower discovered that, at some previous period of the animal's existence, the atlas had been completely dislocated, "the whole of the surfaces, formerly in apposition, being now free from each other," an injury to an aquatic animal as difficult to account for as it is to imagine the possibility of its surviving, but affording a remarkable instance of the creature's recuperative power.

The Whales exhibited at the Westminster Aquarium, in September, 1877, and again in May, 1878, belonged to this species; unfortunately they did not live to equal in docility and intelligence a specimen exhibited in America, which "learned to recognize his keeper, and would allow himself to be handled by him, and at the proper time would come and put his head out of the water to receive the harness" by which he was attached to a car in which he drew a young lady round the tank,—or to take his food. A specimen of *Delphinus tursio*, which was for a time with him in the same tank, is said to have been even more docile than this remarkable animal.[*] The adult Beluga is pure white, and a "school" of these animals "leaping and playing in the calm, dark sea," is said to be a very beautiful sight. In summer the Greenlanders kill great numbers, extracting the oil and drying the flesh for winter use; in Russia, the prepared skin is much used for reins or other parts of

---

[*] Ann. and Mag. Nat. Hist., 3rd series, vol. 17, p, 312.

Fig. 23. THE GRAMPUS (*Orca gladiator*, Lacép.)

harness requiring great strength and lightness; in this country, too, under the name of porpoise-hide, it is now extensively used, and the salted skins sell for from 3s. 6d. to 4s. 6d. per lb. The whale-ship, "Arctic," of Dundee, brought home 600 skins from Davis Strait, in the season of 1880. The length of the full-grown animal is about 16 ft., and its food consists of fishes, Crustacea, and Cephalapods.

THE GRAMPUS, OR KILLER.

The common GRAMPUS, or KILLER (*Orca gladiator*, Lacépède), (fig. 23) is a well-known and widely-dispersed species, being found in both the North Atlantic and Pacific Seas. Andrew Murray says "the common Grampus tumbles through the heavy waves all the way from Britain to Japan, *viâ* the North-west Passage." In the British seas it is frequently met with, and has occurred in several instances on the coast of Norfolk. This species is very fierce, its appetite insatiable, and carnivorous in the strictest sense of the word; to the Greenland and White Whale, as well as to Porpoises and Seals, it is an implacable enemy, and follows them ruthlessly. Dr. Brown says, "the White Whale and Seals often run ashore, in terror of this cetacean, and I have seen Seals spring out of the water when pursued by it. The whalers hate to see it, for its arrival is the signal for every Whale to leave that portion of the ice." Eschricht took out of the stomach of a Killer, 21 ft. long, which came ashore in Jutland, no less than thirteen common porpoises and fourteen Seals.

The rounded, compact form of this species gives the idea of great strength and swiftness, and the beautifully-polished glossy black skin of the back contrasting with the equally pure and well-defined white of the lower parts has a very striking effect; over the eye there is a well-defined white spot. It is a very handsome species, but there is something in its appearance which seems to indicate its cruel nature. Thirteen or fourteen strong, slightly curved teeth are found on either side of both jaws; the flippers are broad and oval-shaped, the dorsal fin high and falcate, particularly in the male.

As my object is mainly that of assisting in the identification of casual visitors to our shores, rather than of giving anything like a history of the known British species of Cetacea, it may be desirable to mention here a very remarkable form, which, although it has never been known

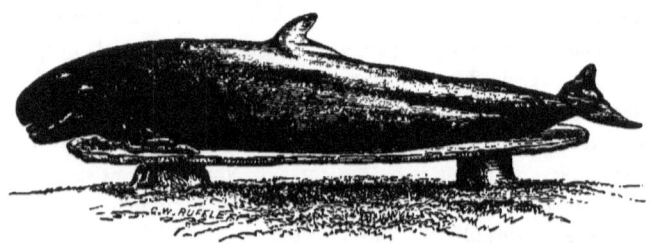

Fig. 24. *Pseudorca crassidens* (Owen).

to occur in the flesh on our shores, was first made known to science from an imperfect skeleton found in a semi-fossil condition beneath the peat in a Lincolnshire Fen. To this Dolphin, "come back, as it were, from the dead," and which forms a connecting link between the genus *Orca* and the genera *Grampus* and *Globicephalus* (and which Owen had named *Phocæna crassidens*), Reinhardt gives the name of *Pseudorca crassidens*. On the 24th November, 1861, a large shoal of these dolphins made their appearance in the

Bay of Kiel. The sailors succeeded in separating about thirty of them from the remainder, but all, with one exception, escaped. This was a female 16 feet long, which, after being exhibited at Kiel and other places, was bought for the Museum of the University of Kiel. In the summer of 1862, three other individuals, presumably from the same shoal, were thrown ashore on the north-western coast of Zealand. Of the general appearance of this creature the accompanying figure (24), copied, by kind permission, from Professor Flower's translation of Reinhardt's paper,* published by the Ray Society, will give an idea; the figure is from a photograph of the Kiel specimen, and is not in the original paper. The length is from 16 to 19 feet; of the colour no account is given, but, judging from the woodcut of the Kiel specimen, it appears to be uniformly shiny black. The number of teeth differs in individuals, but in this one it was from 9 to 10 on either side of the lower jaw, and 8 to 10 in the upper. From the observations made by Reinhardt, he suggests a possibility that there may be "a difference in the sizes of the different sexes, and whether the females are not larger, but at the same time, perhaps, provided with a head comparatively smaller than that of the males." It is very suggestive of how little we know of the inhabitants of the sea, that at least one vast shoal of a species known only from its sub-fossil remains should be roaming the seas only to be accidentally discovered when its members became entangled in shallows from which probably many never lived to extricate themselves.

RISSO'S GRAMPUS.

RISSO'S DOLPHIN (*Grampus griseus*, G. Cuvier; *Grampus cuvieri*, Gray, Ann. Nat. Hist., 1846) is a rare and little-known species, which has been met

---

* Read before the Royal Danish Society of Sciences, in 1862.

with four times on the south coast of England, and about eight times in France. In the 'Transactions' of the Zoological Society, for 1871, Professor Flower gives an account of an adult female which was taken in a mackerel-net, near the Eddystone Lighthouse, on 28th February, 1870, and which eventually was sent up to London. About a month later, a second specimen was received in London, the precise locality of which was not known, but it was probably from somewhere in the Channel. This was also a female, but a very young animal, and as the adult female first taken had recently given birth to a young one, it is quite possible that it may have belonged to her.

Fig. 25. RISSO'S DOLPHIN (*Grampus griseus*, G. Cuv.)

On the 26th July, a male of the same species was captured alive at Sidlesham, near Chichester, and sent to the Brighton Aquarium, where it lived for a few hours only.

Risso's Dolphin varies very considerably in its colouration. The Sidlesham specimen was bluish-black above, and dirty white beneath; in the adult female described by Professor Flower (from whose illustration our figure is, with his permission, copied), "the head and the whole of the body anterior to the dorsal fin was of a lightish grey, variegated with patches of both darker and whiter hue. . . . . . Behind the anterior edge of the dorsal fin the general colour of the surface, including the dorsal and caudal fins, was nearly black, though with a large light patch on the upper part of the side directly above the pudendal orifice. The middle of the belly as far back as the pudendal orifice, was greyish white."* The most remarkable characteristic, however, was the presence, scattered over the body, of irregular light streaks and spots; these markings extended from the head to within about two feet from the tail; and presented a most singular appearance. In the young one the upper parts and sides of the body were almost black, and the lower parts nearly white, the junction between the two colours being very abrupt and sharp. "On each side of the body were six vertical whitish stripes nearly symmetrically arranged, and almost equidistant, being about six inches apart. They did not extend quite to the middle line of the body above, and were lost below in the light colouring of the abdomen."† The length of the Sidlesham male was 8 feet, that of the adult female 10 ft. 6 in.; in the former there were present four teeth on each side the lower jaw, in the latter three only on each side, and in the immature specimen there were present seven teeth, four on the right, and three on the left side; the teeth are always placed in the front part of the mandible, and in every specimen examined there has been an entire absence of teeth in the upper jaw. In general appearance, Risso's Dolphin, more particularly the dark-coloured specimens, is said very much to resemble the next species (*Globicephalus melas*). Of its habits and distribution nothing positive is known, but from its visiting France and England in the spring or summer, M. Fischer concludes that this species "is migratory, visiting the shores of Europe in the summer, and passing in winter either to the south towards the coast of Africa, or to the west towards the American Continent."‡

* Trans. Zool. Soc., vol. viii, p. 3.   † *l. c.* p. 13.   ‡ *l. c.*, p. 18.

THE PILOT WHALE.

The PILOT WHALE (*Globicephalus melas*, Trail; *Delphinus melas*, Trail; *D. globiceps*, Cuv.; *D. deductor*, Scoresby), known in Shetland as the Ca'ing or Driving Whale, is a frequent, although a very uncertain, visitor in British waters. It is met with, according to Lilljeborg, in the North Sea and northern part of the Atlantic Ocean, occasionally as far north as Greenland;

Fig. 26. (PILOT WHALE *Globicephalus melas*, Trail).

off the Orkney and Shetland Islands, and on the North-west coast of Norway, it frequently makes its appearance; and it has been found on the British coast as far south as Cornwall. In Bell's 'British Quadrupeds' it is said that it also appears to enter the Mediterranean. This species is pre-eminently gregarious, and generally occurs in large herds, often numbering several hundreds. So strong is their habit of association that they follow the leading Whale like a flock of sheep, a habit of which the Orkney and Shetland Islanders are fully aware, and avail themselves to the full. When a herd appears in one of the bays, boats immediately put off, and if possible, get to seaward of them, then gradually approaching, with shouts and splashes, they urge the whole herd shoreward, and are generally successful in driving a large number of Whales into shallow water; but should the leader break through the line of boats, the probability is that no efforts the boats' crews can make will prevent all its companions following. Bell gives many instances of large numbers of these animals being taken, the last of which, quoted from the 'Zoologist' for 1846, is, perhaps, the most extraordinary. It is there stated, "on newspaper authority," that 2,080 were taken in Faroe in the previous year within six weeks, and that 1,540 were killed *within two hours* in Quendall Bay, Shetland, on the 22nd September, 1845.

As it too frequently happens that the unfortunate cetaceans which fall into the hands of the fishermen are simply hacked to pieces, and die only from exhaustion arising from loss of blood, it is worthy of remark that, according to Herr Collett, of Christiania, in Norway they are readily killed by a rifle shot, in the throat, or under the breast.

This species (fig. 26) is remarkable for its peculiarly rounded head,—hence its generic name; the flippers are long and pointed, the dorsal fin long and low; the teeth are about an inch in length, seldom all present in the adults, and the normal number, according to Bell, about twenty-four on either side each jaw; ten to twelve is, however, the more usual number present. The length of the adult is about nineteen or twenty feet, its colour glossy black, with the exception of a white stripe along the belly, which has a heart-shaped termination under the throat. Its favourite food is said to be cuttlefish. The figure is copied, with permission, from the 'Transactions' of the Zoological Society, vol. viii., pl. 30.

PORPOISE.

The COMMON PORPOISE (*Phocæna communis*, F. Cuv.; *Delphinus phocæna*, Linn.) is the best known of the Cetacea inhabiting the North Sea, being met with in abundance all round the British Isles, seldom occurring far from land, and often ascending large rivers for a considerable distance: it has been seen in the Thames as high as London Bridge.

Nothing can be more interesting than to watch a shoal of these animals at sea, sometimes tumbling and gambolling under the bows of the vessel which is passing rapidly through the water, with as much ease as if she were motionless, or chasing each other playfully round and round the ship as she lies becalmed, their white bellies glistening in the clear sea, and frequently, apparently out of pure mad delight, leaping completely out of the water, returning to their native element with a most determined header. But it is not till seen in the glass-sided tank of the aquarium that the beauty, and even poetry of motion of these animals can be fully appreciated; swimming along in a series of gentle curves, they just bring the blow-hole to the surface, breathe without stopping, and continue the curve, till in due course they reach the surface again. This is repeated for the whole length of their spacious tank, or is varied by unexpected eccentricities, all indescribably graceful. Under these favourable circumstances for observation it is also clearly seen that the horizontal tail is the propeller which gives the motion; the alternate upward and downward pressure of this organ against the water evidently producing the graceful mode of progression which is so difficult to describe, but so easily understood when witnessed. The flippers are not used as propellers. When the animal is moving forwards they are laid back, against the body; but when it wishes to stop, they are stretched out at right angles to it, so as to offer a resistance to the water, and so arrest the onward motion of the animal. All this, although perfectly understood in theory before, strikes the beholder as a new and beautiful sight when first viewed in practice, from a stand-point, on a level with the animal itself, and as it were in its own element.

The food of the Porpoise consists of fish, and it follows the shoals of herrings, &c., amongst which it commits great depredations; it has a taste for salmon, and is sometimes taken in the salmon-nets. The period of gestation is said to be six months, and it brings forth one young one at a birth; its colour is black on the back, shaded off to silver-grey on the belly, the whole skin beautifully smooth and polished. The teeth number about twenty-five on each side of either jaw, and are spatulate, with a contracted neck, unlike the usually conical teeth of the *Delphinidæ*. The length is four or five feet. The flesh of the Porpoise seems formerly to have been esteemed as an article of food, and is mentioned several times in the L'Estrange Household Book (1519 to 1578) and other similar records; it is said by one who has eaten it to be "excellent meat, dark in colour, and large in fibre, but of excellent flavour, very tender, and full of gravy."

THE COMMON DOLPHIN.

The COMMON DOLPHIN (*Delphinus delphis*, Linn.), fig. 27, is not unfrequently met with in the seas surrounding the southern portion of the British Isles; but from the northern division of the kingdom, although it, doubtless, occasionally visits Scottish waters, there is no reliable record of its occurrence. This species, probably, often passes unrecognized. It may, however, be at once distinguished from the Porpoise by its attenuated beak, the head of the

Porpoise being obtuse, and the beak altogether absent. It is a native of the temperate seas, and becomes scarcer as the north is approached. Van Beneden was not able to record it as frequenting the Belgian coast, but Lilljeborg says it is occasionally obtained on the coasts of Scandinavia, and Herr Collett has hardly any doubt that it occurs on the Norwegian coast as far north as Finmarken, and a large "school," seen by Malmgren in April, 1861, in West-fjord, between the Loffoden Islands and the mainland, was referred by him, without hesitation, to this species. In Greenland it is said to be met

Fig. 27. COMMON DOLPHIN (*Delphinus delphis*, Linn.).

with, but Professor Flower thinks it doubtful whether some species of an allied genus may not have been mistaken for it.

This is the true Dolphin of the Ancients, of which Professor Bell, in his 'British Quadrupeds,' says: "the mythological and poetical associations which belong to the Dolphin, its reputed attachment to mankind, its benevolent aid in cases of shipwreck, its dedication to the gods, and many other attributes expressive of the high estimation in which it was held in olden times, afford a striking example of how the unrestrained imagination of the ancients could raise the most gorgeous structures of poetry and religion upon the most slender basis . . . . It requires some stretch of the imagination to identify the round-headed creature which is represented in ancient coins and statues, with the straight sharp-beaked animal," which is here figured. It is sad to destroy at one fell swoop all the romance which once surrounded this species; but Dr. Gray tells us that "the dying Dolphin's changing hues" are not observed in a cetacean at all, but in a fish of the genus *Coryphæna*, which, although normally black, is stated by Mr. Couch (as quoted by Mr. Yarrell) to have changed to a fine blue whilst he was making a drawing of it. The food of the Dolphin consists of fish, cuttlefish, and crustaceans, and on the Cornish coast it makes its appearance in considerable numbers, according to Mr. Couch, in the month of September during the pilchard season. It is very social in its habits, and even more sportive in the water than its relative, the Porpoise. The illustration is copied from Reinhardt's figure.

Professor Flower thus describes a specimen taken in March, 1879, at Mevagessey: "Instead of being simply black above and white below, as usually described, the sides were shaded, mottled, and streaked with various tints of yellow and grey, ... the under surface was of the purest possible white; perfect symmetry was shown in the colouring and markings on the two sides of the body."* There is, probably, much variation in the disposal of the colour; in a beautiful drawing, in my possession, made by Mr. Gatcombe from a specimen taken at Plymouth, the colour is so disposed as to show two graceful waving lines, crossing each other about the centre of the animal's body, forming a figure somewhat like an elongated figure eight. The dental formulæ vary from $\frac{40}{40} \frac{40}{40}$ to $\frac{50}{50} \frac{50}{50}$, the numbers not always being equal, even on the different sides of the mouth of the same individual. The length is from 5 to 8 feet.

* *Trans. Zool. Soc.*, vol. xi., p. 2, with plate.

BOTTLE-NOSED DOLPHIN.

The BOTTLE-NOSED DOLPHIN (*Delphinus tursio*, Fab.; *Tursio truncatus*, Gray), fig. 28, appears to be found occasionally from the Mediterranean to the North Sea; it is by no means, however, a common species. Professor Flower says it "is rare in the Mediterranean, though Gervais gives several instances

Fig. 28. BOTTLE-NOSED DOLPHIN (*Delphinus tursio*, Fabricius.)

of its capture in the Gulf of Lyons. It probably has a more northern range than *D. delphis*; but, as in the case of that species, there is still much obscurity as to the exact limits of its distribution."* A specimen was seen in January, 1873, in the fishmarket at Algiers, by Mr. J. W. Clark, of Cambridge.

---

* *Trans. Zool. Soc.*, vol. xi., p. 5.

Of the habits of this species very little is known: its colour is black above, shaded to white below, and its length from 8 to 12 feet; teeth from 21 to 25 on either side of each jaw, truncated when old. The figure is from a drawing of a nearly adult male, taken at Holyhead, in October, 1868, for which I am indebted to the kindness of Professor Flower.

WHITE-SIDED DOLPHIN.

The WHITE-SIDED DOLPHIN (*Delphinus acutus*, J. E. Gray; *Lagenorhynchus acutus*, Gray, Zool. Erebus and Terror), is a rare species, which has occurred in a few instances on the British coast; it is said, however, by Dr. A. R. Duguid, often to be seen about the Orkney Islands, but rarely secured. Its colour is black above and white below, between which runs a broad band of yellowish brown, about the centre of which, and surrounded by it, is a large oblong patch of pure white. The adult measures from 6 to 8 feet in length. A figure and description, by Dr. Duguid, taken from one of a herd of twenty landed at Kirkwall, on the 21st August, 1858, will be found in the 'Ann. and Mag. of Nat. Hist.' (3rd series) for August, 1864, vol. xiv., p. 133.

WHITE-BEAKED DOLPHIN.

The last species on the British list, the WHITE-BEAKED DOLPHIN (*Delphinus albirostris*, J. E. Gray; *Lagenorhynchus albirostris*, J. E. Gray, Zool. Erebus and Terror), is also of rare occurrence: it is a native of the North Atlantic, has occurred at the Faroe Islands, and on the coasts of Norway and Sweden, and Denmark, also at Ostend, but little is known of its habits. A Dolphin of this species was killed at Hartlepool in 1834, but not recognized at the time: the skull is now in the Cambridge Museum. This species was, I believe, first described as British by Mr. Brightwell, under the name of *D. tursio*, from a specimen taken off Yarmouth, in 1846. His paper, with a figure from a drawing made by Miss Brightwell, will be found in the 'Ann. and Mag. of Nat. Hist.,' first series, January, 1846, vol. xvii. p. 21. Another specimen was

Fig. 29. WHITE-BEAKED DOLPHIN (*Delphinus alberostris*, J. E. Gray).

shot by Mr. H. M. Upcher, near Cromer, and will be found recorded by Dr. Gray in the same Magazine, for April, 1866, vol. xvii., p. 312. A fourth, an adult male, 9 feet long, was taken at the mouth of the Dee, in December, 1862; and a fifth on the south coast, in 1871.

In September, 1875, a young female was taken off Grimsby, and in March, 1876, a young male was captured off Lowestoft. The first-named of these latter formed the subject of a communication to the Zoological Society of London, by Dr. Cunningham, of Edinburgh, and the latter of a subsequent notice, by Mr. J. W. Clark, of Cambridge. Both papers will be found printed in the 'Proceedings' of the Zoological Society for 1876, p. 679, *et seq.*, and figures of the two specimens are given on the same plate. On the 24th August, 1879, a young female, the skull of which is now in the Norfolk and Norwich Museum, was landed at Yarmouth, and on the 22nd March, 1880, another young female was also landed at the same place, the exact locality in which it was taken being uncertain. On the 7th September, 1880, a young male, the first recorded Scotch specimen, was taken on the east coast, near the Bell Rock, thus realising the belief, expressed shortly before ('Mammalia of Scotland,' *Nat. Hist. Soc. of Glasgow*, 1880, p. 23) by Mr. Alston, that it might be expected to occur in Scottish waters. The total length was 5ft. 8 in.

Through the kindness of Mr. Clark, I am enabled to give a figure of the Lowestoft specimen. Mr. Clark's figure differs considerably from Dr. Cunningham's, both in outline and in the disposal of colour, being much more slender, and showing considerably less white; both, however, differ still more from Mr. Brightwell's figure than they do from each other. A good figure of the adult animal is still a desideratum, that by Miss Brightwell being obviously incorrect. Mr. Clark's specimen was glossy black on the upper part, and creamy white on the under; the upper lip white, with a black spot at the tip, and a few irregular pale grey cloudings on its surface; the coloration exceedingly beautiful, and such as no drawing could give an adequate idea of. The two last-named Yarmouth examples agreed very closely in all respects with Mr. Clark's description. Mr. Brightwell's specimen had the whole upper part and sides rich purple-black, the lips, throat, and belly cream-colour, varied by chalky-white. This specimen, an adult, measured 8 ft. 2 in. in length, Mr. Clark's 5 ft. 5½ in., and Dr. Cunningham's 4 ft. 2 in. Two others, also both young ones, measured respectively 4 ft. 3 in., and 5 ft. The teeth vary in number, but are about twenty-six on either side each jaw;

in one specimen, carefully examined by the writer, they were $\frac{24}{24}$ $\frac{24}{24}$, several of the front teeth not having pierced the gum.

In addition to those enumerated above, others are said to have occurred on the coast of Belgium, Denmark, Norway, Sweden, and to have been seen off the Färoe Islands. It is singular that 5 of the 10 recorded British specimens should have been landed on the Norfolk coast.

This species concludes the short list of the twenty-two British Cetacea, of which I have endeavoured to give a popular, but I hope, at the same time, so far as it is at present known, a reliable account; my principal object, as I stated in my introductory remarks, being to induce those residing in suitable localities to take up the study of this interesting family, and to assist in the identification of those specimens which from time to time are cast upon our shores.

NOTE TO PAGE 77, RUDOLPHI'S RORQUAL (*Balænoptera laticeps*, J. E. Gray).—Professor Flower, since the brief account of this animal at p. 77 was printed, has called my attention to the undoubted priority of Lesson's name for this species, *Balænoptera borealis*, which was founded upon Cuvier's "Rorqual du Nord"; he also points out that Van Beneden and Gervais follow Lesson in this respect, and says that in future it is his intention to do the same. As it is most important to establish an uniform nomenclature, I do not hesitate to follow so distinguished an authority, and now wish to supply the omission as far as it is possible to do so. The species will, doubtless, henceforth be known as *Balænoptera borealis*, Lesson, Complément des Œuvres de Buffon, Cetacés.

JARROLD AND SONS, PRINTERS, LONDON AND EXCHANGE STREETS, NORWICH.

Cloth, 6s.; or in Half Morocco, 10s. 6d.

OBSERVATIONS ON THE FAUNA OF NORFOLK,

AND MORE PARTICULARLY ON

THE DISTRICT OF THE BROADS.

BY

THE LATE REV. RICHARD LUBBOCK, M.A.,

Rector of Eccles.

NEW EDITION,

WITH ADDITIONS FROM UNPUBLISHED MANUSCRIPTS OF THE AUTHOR, AND NOTES BY

THOMAS SOUTHWELL, F.Z.S.,

Hon. Sec. to the Norfolk and Norwich Naturalists' Society; Author of "Seals & Whales of the British Seas;"

ALSO A MEMOIR BY

HENRY STEVENSON, F.L.S.;

AND AN APPENDIX CONTAINING NOTES ON HAWKING IN NORFOLK BY

ALFRED NEWTON, M.A., F.R.S., &c.

AND ON THE DECOYS, REPTILES, SEA FISH, LEPIDOPTERA, AND BOTANY OF THE COUNTY.

OPINIONS OF THE PRESS.

"Lubbock's volume, written five-and-thirty years ago, has long been out of print and scarce; and the reliable nature of the information which it affords has for some time rendered a new edition a *desideratum* with naturalists. A new edition has at length appeared, edited by Mr. Thomas Southwell, of Norwich, who has made some valuable additions of his own in the shape of notes on the existing mammalia of Norfolk, and on decoys past and present in the county, prefaced by a memoir of the author by Mr. Henry Stevenson, and supplemented by some interesting notes on Hawking in Norfolk, from the pen of Professor Newton."—*The Field.*

"In addition to the intrinsic merits of the book, of which we can personally speak in the superlative degree as one of the most pleasantly written of the many pleasant natural history books our language is so rich in, describing, as it does, the 'Broad District' —a country unlike any other part of England, and a very paradise to the botanist, entomologist, and ornithologist—this new edition is edited by Mr. Thomas Southwell, the active Secretary of the Norfolk and Norwich Naturalists' Society, whose full and accurate knowledge of the natural history of Norfolk better fits him for the task than any other man we know of."—*Science Gossip.*

"While Mr. Lubbock's personal observations were chiefly directed to the neighbourhood of the Broads, the editor has endeavoured to make the work as comprehensive in its scope as possible, and he includes the district known as Lothingland, between Lowestoft and Yarmouth, which, though in Suffolk, belongs geographically to Norfolk."—*Midland Naturalist.*

"We promise to those who have never yet read this book, a rare treat from its perusal."—*Zoologist.*

"We can scarcely speak too highly of the way in which this volume has been 'got up,' and the publishers have added such a map as has never yet been executed of this county, showing, as it does, not only the rivers and broads, and other principal pieces of water, but the sites of heronries and decoys (used or disused), gulleries, and other localities, having a special interest for Naturalists."—*Norfolk Chronicle.*

"The 'Fauna' is a book which everyone should read who desires to know something of the natural history of Norfolk."—*Norfolk News.*

"Absolutely reliable and authoritative as a work of reference, and invaluable to every naturalist and ornithologist."—*Live Stock Journal.*

JARROLD AND SONS, 3, PATERNOSTER BUILDINGS, LONDON;
AND LONDON AND EXCHANGE STREETS, NORWICH.

Large 8vo., Cloth Boards, Seven Shillings and Sixpence.

Rambles of a Naturalist

IN

EGYPT AND OTHER COUNTRIES,

WITH AN ANALYSIS OF THE CLAIMS OF CERTAIN
FOREIGN BIRDS TO BE CONSIDERED BRITISH, AND OTHER
ORNITHOLOGICAL NOTES.

BY J. H. GURNEY, Jun., F.Z.S.

JARROLD AND SONS, 3, PATERNOSTER BUILDINGS, LONDON;
AND LONDON AND EXCHANGE STREETS, NORWICH.

www.ingramcontent.com/pod-product-compliance
Lightning Source LLC
Chambersburg PA
CBHW030900170426
43193CB00009BA/689